UK AND EUROPEAN SCIENCE POLICY

The Policy Studies Institute (PSI) is Britain's leading independent research organisation undertaking studies of economic, industrial and social policy, and the workings of political institutions.

PSI is a registered charity, run on a non-profit basis, and is not associated with any political party, pressure group or commercial interest.

PSI attaches great importance to covering a wide range of subject areas with its multidisciplinary approach. The Institute's researchers are organised in groups which currently cover the following programmes:

Crime, Justice and Youth Studies – Employment and Society – Ethnic Equality and Diversity – European Industrial Development – Family Finances – Information and Citizenship – Information and Cultural Studies – Social Care and Health Studies – Work, Benefits and Social Participation

This publication arises from the European Industrial Development group and is one of over 30 publications made available by the Institute each year.

Information about the work of PSI, and a catalogue of available books can be obtained from:

Marketing Department, PSI
100 Park Village East, London NW1 3SR

UK and European Science Policy

The role of collaborative research

Kirsty Hughes with Ian Christie

POLICY STUDIES INSTITUTE
London

The publishing imprint of the independent
POLICY STUDIES INSTITUTE
100 Park Village East, London NW1 3SR
Telephone: 0171-387 2171 Fax: 0171-388 0914

ISBN 0 85374 580 3

PSI Research Report 795

A CIP catalogue record of this book is available from the British Library.

1 2 3 4 5 6 7 8 9

PSI publications are available from:
BEBC Distribution Ltd
P O Box 1496, Poole, Dorset, BH12 3YD

Books will normally be despatched within 24 hours. Cheques should be made payable to BEBC Distribution Ltd.

Credit card and telephone/fax orders may be placed on the following freephone numbers:

FREEPHONE: 0800 262260
FREEFAX: 0800 262266

Booktrade representation (UK & Eire):
Broadcast Books
24 De Montfort Road, London SW16 1LW
Telephone: 081-677 5129

PSI subscriptions are available from PSI's subscription agent:
Carfax Publishing Company Ltd
P O Box 25, Abingdon, Oxford OX14 3UE

Laserset by Policy Studies Institute
Printed and bound in Great Britain by Latimer Trend and Co Ltd

Contents

Acknowledgements

The research reported in this book was funded by the Economic and Social Research Council – research grant no. L 32325 3008 – as part of its Research Programme entitled 'The European Context of UK Science Policy' managed by the Science Policy Support Group. Views expressed are those of the authors alone. We are grateful to the many participants in the case studies for the time they took to discuss their work with us. We are also grateful to Karin Erskine and Esther Lane for their work in preparing the manuscript.

1 Collaborative Research: Motivations, Implications and the Role of Policy

This study is concerned with the analysis of co-operative scientific research networks. It focuses in particular on co-operation between scientists in universities and research institutes rather than on inter-firm collaboration. As scientific collaboration in networks becomes increasingly common, it is important to ask what the motivations are for participation in these networks, what the implications and outcomes of science taking place in such networks are, how different forms of networks compare to each other, and how networks compare to alternative models and structures for research. This study addresses these questions.

The sharp increase in collaboration in research during the 1980s affected scientists both in universities and firms. These developments have implications for, and are influenced by, science policy and industrial policy, and there is a significant blurring of the boundaries between science and industrial policy that needs to be given serious attention. Given that much of the collaboration is international, in the sense of collaboration between organisations in Europe, the US and Japan, questions of appropriate policy go beyond those of the balance between EU policy and national science policy – the role of subsidiarity – to questions of what sort of international co-operation and policy instruments are relevant.

This chapter assesses the motivations for and implications of collaborative research and considers some of the policy issues. On the basis of this analysis, an analytical framework for investigating co-operative scientific research networks is established in chapter two. The subsequent chapters then present the results of eight case studies into European and UK scientific networks, and the final chapter analyses the implications and main results from the case studies both for our understanding of co-operative research and for policy.

Types of collaborative research activities
There is a wide variety of forms of collaborative research activities in terms of the participants, the structure of the collaboration, and the aims of the collaboration. Collaborative research may take place solely between firms; it may take place solely between scientists in universities or research institutes; it may

1

take place between industry and universities; and it may include government research establishments. Collaboration may take the form of a tightly specified joint venture between two companies or it may take the form of a loose network between a large number of scientists. In between, there is a wide range of possible structures. Collaborative agreements between firms may focus only on research or they may also include production and marketing agreements. The bulk of the literature focuses on either industry research collaboration or industry-university research collaboration. However, there may be important differences between networks dominated by firms and market motivations and networks where scientists in not-for-profit institutions are participating.

Research collaboration and research networks can be seen as alternative forms of organisation, or alternative forms of relationships between actors, to markets and to hierarchies or fixed organisational structures. As such, they can encompass a wide range of structures and behaviour. Networks and collaboration can provide an extremely valuable balance between commitment and flexibility that neither markets nor hierarchies provide (Jacquemin 1988, OECD 1992).

During the 1980s, there was a substantial increase in collaborative research, in particular in international research joint ventures. This increase in collaboration has been observed both within the three major blocs – Western Europe, Japan and the US – and across the three blocs (Ohmae 1985). This can partly be seen, especially in Japan and Europe, as linked to policy developments but this collaboration has also been at the instigation of firms. There is considerable information on the number and area of these joint ventures but less information about their precise structures, functioning and outcomes. This more detailed information usually requires a case study approach. In an analysis of over seven thousand joint ventures, Hagedoorn and Schakenraad (1993) found that there was an increase in the number of joint agreements in the second half of the 1980s. The largest number of agreements were in the area of information technology – over 40 per cent – and the next largest was in bio-technology – about 20 per cent. The largest share in these agreements were intra-US agreements followed by agreements between European companies and the US followed by intra-European agreements; US-Japan and Japan-Europe agreements were smaller in number. It appears that in the second half of the 1980s intra-bloc agreements increased relatively. It is argued that US and Japanese companies are more globally oriented in their technological alliances than are European companies.

Motivations for collaboration
There are a variety of reasons why firms and other organisations such as universities collaborate in research. These reasons are to do with the nature of the research and R&D process but there are also various motivations that go beyond this to concerns with markets or other possible benefits of collaboration. Some of the major reasons put forward for collaborative research include: reduction in costs and risks, increase in appropriability, access to expertise and

knowledge – especially tacit knowledge and know-how – and externality benefits such as collusion down-stream in product markets, market access and long-run contacts that may be used for other purposes in the future.

Joint research will reduce the cost of research if there are scale economies involved. In addition it will potentially act to reduce risks associated with innovation. The risks associated with innovation can be classified according to three categories – demand uncertainty, technological uncertainty, and uncertainty about behaviour and likely success of competitors. Joint research spreads these risks across firms and at the same time also alters and reduces the nature of uncertainty about competitors' actions. The firms that may have been previously duplicating research will be able to benefit from undertaking a joint programme which will reduce cost to all participants. Alternatively, the cessation of duplication may allow firms to broaden their range of activities and so increase their chances of success in innovation (Grossman & Shapiro 1987, Dasgupta & Maskin 1987). However, whether collaborative research can reduce or eliminate market failures is likely to depend on a number of factors including on whether the co-operative venture includes all competing firms (see, for example, Ordover & Willig 1985, Katz 1986, and Cohen 1994).

One particular problem raised by competition in R&D is the difficulty of appropriating R&D output. Through joint R&D, spill-overs from firms' R&D can be internalised (Katz 1986, Geroski 1992). Katz & Ordover (1990) acknowledge that while the social rate of return to innovation is generally greater than the private rate of return, firms can in certain circumstances do too much R&D. At the same time, co-operation can decrease the incentive to do R&D. Research joint ventures may reduce the incentive to innovate by reducing the competitive stimulus between firms. Subsidies for collaboration can also in certain circumstances reduce the incentive to innovate (Cohen 1994). Joint research does not solve all the appropriability problems associated with innovation. Furthermore, while co-operation may avoid profit dissipation through R&D rivalry, a research joint venture may increase direct rivalry in the market and so reduce profits (Grossman & Shapiro 1987).

Much of the analysis of collaborative research discussed so far does not pay particular attention to differences between firms or other collaborating parties. Once differences between organisations are acknowledged then a wider range of reasons for collaboration can be identified. These differences are likely to include differences in expertise, knowledge and information. There are well known difficulties in trading information and knowledge, especially where knowledge is tacit; collaborative ventures may be an effective way of allowing the exchange of such knowledge. Differences between collaborators may of course be asymmetric – one firm may have superior technological knowledge to another but still benefit from access to the weaker firm's markets, productive capacity, scientific and technical skills and so forth. Questions then arise about who is getting most benefit from the collaboration and whether any party to a collaboration may lose and what can be done about this. Equally, genuine

3

complementarity among partners can encourage successful collaboration (Sharp & Shearman 1987). Overall, where there are differences between partners, R&D co-operation may be seen to be preferable to licensing or cross-licensing due to problems of moral hazard, opportunism and due to the tacit, uncodified nature of parts of technological know-how and the difficulties of separating this know-how from the rest of the firm when it may be closely dependent on particular staff and particular organisational structures (see, for example, Teece 1981). Furthermore, costs of transferring technology can be very high (Teece 1977, Rosenberg & Frischtak 1985).

Collaboration between firms takes many forms, not only formal joint ventures but also quite informal contacts. Informal know-how trading among rival firms has been observed in some sectors (von Hippel, 1987). Conditions which encourage informal know-how trading, where there are no formal agreements, include reciprocity and repetition. Repeated interaction allows the dangers of unequal exchange that might otherwise occur to be overcome. Informal networks may also be important for the success of formal collaboration (Dickson et al 1991).

Collaborative R&D or collaborative research networks may facilitate the transfer of knowledge and information but there may still be problems in resolving the difficulties of information and knowledge transfer. Transfer of tacit knowledge may be easier through collaboration than through the market and easier to control but it can still encounter difficulties and will not be costless (Mowery 1987). Distinctions may also be drawn between knowledge and information transfer, with knowledge transfer requiring more complex mechanisms. Reciprocity and trust are seen by some as critical ingredients of successful knowledge transfer (OECD 1992). Networks may be fairly loose structures but joint research collaboration is usually aimed at the creation of new knowledge and innovation and not simply at the transfer of existing knowledge. At some point this knowledge or output has to be transferred back and used by the participating bodies. This transfer is not likely to be costless and it also raises questions about the relationship between the participants when they use the output of the collaborations subsequent to the collaboration. To the extent that organisations develop rigidities in their patterns of information and accumulation, co-operation may break these down and create new synergies (Imai & Itami 1984). Collaboration is not costless – in addition to technology transfer costs, there are co-ordination costs and potential costs due to diversion of research funds and potential loss of competitive advantage (Link & Tassey 1989).

Causes of success and failure in collaborations and networks are not easy to identify. Hingel (1993) comments on the lack of knowledge about what actually goes on within networks. At least in the 1980s, many research collaborations could be said to have been experimental and so could not all have been expected to succeed (Nelkin & Nelson 1986). Ohmae (1985) and Mowery (1987) both provide interesting discussions of causes of success and failure in joint ventures of various sorts. Generalisation is difficult, but the importance of effective communication comes out strongly, as does the importance of effective management and the need for clarity and transparency in management and organisational structures. Substantial learning occurs in processes of co-operation, both within a particular joint venture and through repeated experience (de Meyer 1992).

Competition

There has been substantial discussion of the potential competition implications of collaborative research among firms, with rather different views being taken. Some see collaborative research as likely to lead to and encourage collaboration in product markets; others see this as unlikely or even that collaboration and research can intensify subsequent product market competition. Ohmae (1985) argues that the increasing scale of R&D and the need for co-operative R&D may indeed require such links to be extended to production and marketing in order to gain access to global markets to recover the high costs of R&D. International joint ventures can provide the necessary knowledge of markets and the appropriate marketing and distribution skills. Competition can still come from domestic competitors and other international joint ventures. Baumol (1992) argues that collusion can increase prices and even lead to the avoidance of innovation. However, collusion may be acceptable if it results in higher productivity and growth and lower costs. Even if collaboration results in lower innovation, the overall benefits can be positive due to lower costs and higher and faster dissemination of the results of innovation. He argues that the importance of dissemination has been underestimated. Jacquemin (1988) also argues that the benefits of co-operation outweigh the costs sufficiently to demand that competition regulations take into account the beneficial aspects of research collaboration. Vickers (1985) and Katz and Ordover (1990) show collaboration may reduce competition. Overall, there is insufficient empirical evidence on the competition and competitiveness effects of collaboration (Chesnais 1988). In particular, the relative balance of power between firms needs further investigation together with the role played by key 'nodal' firms who are often at the centre of patterns of collaboration and networking (Chesnais 1988). Key nodal players have also been observed in scientists' interactions, with a relatively small number of leading scientists playing a key role in information flows (Crane 1972). If a small number of firms are playing a nodal role, the technological, competition and network implications need further consideration.

While much attention has therefore been given to the implications for product market competition of collaborative research ventures, less attention has been paid to the overall effects of collaborative research on technological competition itself. Some of the authors discussed have shown that collaboration could in some circumstances lead to less R&D than in the absence of collaboration. However, the overall implications of this, if it is a wide-spread trend, are not drawn out. It is possible – as there is increasing collaboration on an international scale, as technological knowledge becomes increasingly common and diffuses increasingly swiftly, and as product life cycles shorten – that competition between firms comes increasingly to rest on factors other than innovation and technology (Hughes 1989). It may be asked whether and to what extent a process of technological levelling is occurring.

The increasing trend in collaborative R&D
Various reasons have been put forward to explain the sharp increase in international joint ventures, particularly in research, in the 1980s and 1990s. Main reasons include: increased intensity of international competition, increased internationalisation of firms and markets, increased 'knowledge-intensity' of production, higher costs of undertaking research, and increased interdisciplinarity and complexity of the research process (see for example, OECD 1992, Fusfeld & Haklisch 1985, Ohmae 1985, Rosenberg 1982, Shepherd et al 1983, Mytelka 1990). Innovation is one element of competition in many markets and products, though its importance varies across markets and sectors. Increased intensity of competition can therefore lead to increased pressure for innovation. Higher innovation pressures may also reflect increased competition from the newly industrialising countries in less technology-intensive products and in some of the technology-intensive areas. Advanced industrial economies may need to focus more on the medium and high technology industries if they are to maintain a competitive advantage. A related issue here is the extent to which countries are catching up and converging with, or in some areas overtaking, the US in terms of technology and productivity. As economies become more similar, there is less scope for imitation and catch-up and so more emphasis may be placed on innovation.

Japan's success in catching up with the US is well documented (see, for example, Freeman 1987). Meeting the challenge from Japan is frequently cited as a reason for R&D co-operation in the US and catching up with both the US and Japan as a reason for European R&D co-operation (see, for example, Shepherd et al 1983, Geroski & Jacquemin 1985). However, as Patal and Pavitt (1987, 1991) have stressed, generalisations about the relative technological position of the US, Japan and West Europe are likely to be misleading. Furthermore, it is inappropriate to treat Western Europe as a homogenous technological entity.

Intensified global competition may lead to increased R&D collaboration in order both to increase the speed of innovation and to reduce the costs and risks associated with a high innovation strategy. These factors may be reinforced by the shortening of product life cycles. Collaboration may also be a reaction where firms or countries perceive themselves to be lagging technologically.

There is a common view that the costs of innovation have increased substantially, though this is not always well substantiated. Nevertheless, in some areas, costs and/or scale of research have risen. At the same time there is an increasing interdisciplinarity in science and technology, an increasingly cross-industry nature of science and technology and a tendency for basic science to have much more immediate relevance to product and process innovation. Increasing complexity and interdependence of technology not only raises costs but also implies that any one firm is unlikely to have the abilities and knowledge necessary for R&D and product and process development. Thus, the evolution of the international technological structure may be such that corporations need to draw on a growing range of technical areas (Fusfeld & Haklisch 1985). Firms are thus increasingly less self-sufficient and are more dependent on external links. As has already been argued above, this may in the medium run have major implications for the nature of technological competition and which other factors become important in international competition.

A number of writers have expressed concern that joint ventures in R&D can result in unequal benefits to the partners. R&D joint ventures may not only concern R&D. Whether this is acknowledged formally or informally, joint ventures may result in access to markets and to other aspects of the firm's mode of operation. This potential inequality in outcomes and motivations in participating in collaborative research is of importance not only to participants and to the likely success of the venture but it can also have implications at the level of the economy as a whole. In particular, in the US some strong views have been expressed that US joint ventures with Japan are giving away the US technological advantage and facilitating access to markets (Reich and Mankin 1986). Reich and Mankin analyse a number of joint ventures between US and Japanese firms and conclude that these have tended to involve transfer of knowledge and learning experience from the US to Japan. Companies should be alert to the dangers of being 'hollowed out' (Lei and Slocum 1992). It is often the case in such arguments that policy conclusions are drawn to the effect that a particular country could act to ensure that it starts to win the technological game and does not lose technological leadership. However, it is clear that not all countries or all firms can win the technological game. Furthermore, the costs of participating in the game have increased substantially. As Nelson (1984) states: 'in the nature of the case, five countries cannot all be first in the product cycle race, and the competition appears to have reduced the size of the prize and increased the cost of entry'.

UK and European science and technology policy
In analysing processes of collaborative research and co-operative research networks, it is important to consider the role of government policy both in influencing and reacting to developments. In the EU context, it is also necessary to consider the interactions and relationship between national science policy and EU science policy. Direct questions can be asked about whether national and EU science policies duplicate each other, complement each other or contradict each other. However, probably the questions of greater interest are how these policies influence each other indirectly through their influence on the behaviour of actors concerned with, and operating within, the context of these science and technology

7

policies. One major study has already been carried out looking at the impact of EU policies on national science policies, including a report for the UK (IMPACT 1993).

UK and EU science policies are in many ways surprisingly similar in their overall aims and motivations – surprising given the different overall approaches of the UK government and the European Commission. What the two sets of policies have in common is that their main aim is to utilise science and technology in such a way as to improve competitiveness. A subsidiary aim, given this overall aim, is to improve the effectiveness of the underlying science base. Both the UK and EU policies can be described as technocratic (Edgerton & Hughes 1993). This technocratic position sees science, technology and innovation as *the* key determinants of a nation's economic performance, growth and international competitiveness. Despite the uncertainties involved inherently in innovation, this approach considers it is possible to predict the main paths of technology and to utilise government action to control and direct the level and direction of innovation towards those key technologies necessary for successful performance in coming years.

This technocratic approach may be seen as somewhat surprising in both the UK and the EU's case as both emphasise the importance of the free market, the latter through its Single European Market Programme. Within the EU, this apparent contradiction between free market and technocratic beliefs can be seen as reflecting a continuing debate both within and outside of the Commission itself, between advocates of a free market and advocates of greater government intervention in industrial policy (Hughes 1992, Sharp 1991). In the UK case, there is no current government policy stance in favour of direct intervention in industry. However, when it comes to science and technology policy, given that government funds major amounts of basic science through the universities, then government sees a role to control and direct research so that it serves the needs of the markets – in this case industry. Thus, the ironic position is reached, that although there is no industrial policy because it is seen as undesirable and impossible to outguess the market, in the case of the much more uncertain area of science and technology advance there is indeed an industrial policy: science policy becomes a de facto industrial policy.

Subsidiarity

While UK and EU science policy have similar overall aims, there are important questions to be asked concerning the relationship between national and EU policies and in particular concerning the issue of subsidiarity. The European Commission itself has said that it wishes to make the Fourth Framework Programme an instrument for a better co-ordination of national and EU efforts (House of Lords 1993a). The relative roles of national and EU science policy are in fact not completely clear. The main rationale for EU science policies is seen to lie in the existence of scale economies, however, once other motivations are included, the separation of responsibilities and so the application of subsidiarity is less clear.

The UK government view of subsidiarity with respect to science policy is essentially one that sees the key area for the EU as being large projects where there are important scale economies. This may include research that requires the use of large expensive equipment and projects involving generic technology with potential multi-sectoral impact where again scale is important (House of Lords 1993a, Science Policy White Paper 1993). One report (House of Lords 1990) identified four areas where there was scope for EU science policy to take precedence. All these four areas can in fact be seen as aspects of the scale economy criteria. They are: development of common standards; projects dealing with EU or international problems such as the environment; projects too costly for one country; and projects where there is added value to undertake them at EU level. With respect to the latter criteria, its interpretation was seen as unclear but to do with the creation of a critical mass of expertise and knowledge that would not be possible in one country.

The European Commission has also emphasised the importance of scale economies in its programmes (CEC 1992). It saw the Second Framework Programme as demonstrating the economies of scale that could be achieved at EU level in research and technological development. Different organisations had seen the benefits of sharing risks and rewards on large-scale projects. At the same time, the Commission saw that benefits had been achieved not solely from scale but also due to the access to other sources of knowledge, skills and equipment. Commenting on the 10 years experience of Framework Programmes in 1993, Professor Ruberti, Vice-President of the European Commission, argued that there were four main achievements of these programmes: the establishment of networks of co-operation; the achievement of important scientific results; industrial impact with respect to industrial technologies; and a quantative impact with respect to the numbers of firms and universities that had participated (House of Lords 1993a). At the heart of the EU's science policy is an emphasis on collaboration and networks. Such an emphasis can help to justify the existence of an EU science policy in addition to national science policies.

Two issues make the appropriate interpretation of subsidiarity in the context of science and technology policy particularly difficult. The first is the fact that benefits from collaboration come not only from scale economies but also from scope economies i.e. from complementarity and diversity. The second is that there is an international science community and scientists interact internationally and globally. These two issues are interconnected. The EU's focus on collaboration and networks means that to some extent it has side-stepped the discussion about the relative role of scale and scope economies and so the interpretation of subsidiarity. However, as the case studies presented in this report suggest, the role of scope economies and diversity may in fact be more important than the issue of scale economies. The importance of scope economies does not mean that collaboration and science policy should not take place at EU level, but it does mean there is a much less clear dividing line between the appropriate focus of national and EU science policy.

This difficulty in distinguishing the relative roles of EU and national science policy is compounded by the international nature of science. Interpreted strictly, the UK government view would imply, that unless there are major scale economies, UK scientists should operate within national boundaries and collaborate nationally. However, if scientists are operating as part of an international community, it would seem inappropriate to try to constrain them to collaborating on research projects only within their own country. This is not simply a question of whether appropriate expertise exists within one country, it is a question of recognising the nature of scientific interaction. Scientists get to know each other and each other's work through papers, results and meetings of various sorts. On the basis both of assessment of each other's work and on the basis of personal interaction, scientists may decide to work together. Given the ease with which scientists can now collaborate internationally due to advanced telecommunications and transport, there are much fewer constraints than previously on this international interaction. If scientists are to use their knowledge about each other and their discipline most effectively, then free selection of partners for collaboration appears quite important (Leclerc et al 1992).

The existence of an international community of scientists raises difficult questions both for national and for EU science policy. Both these policies may to some extent be operating at least some of the time to constrain or influence scientists inappropriately in their choice of research area and partners. Nor are the implications for subsidiarity straightforward. The fact that scientists should be encouraged to interact internationally, which will include on a European level, does not necessarily imply that EU policies should be responsible for this interaction.

EU policies for science stress collaboration and scale. Once the importance of international interaction and of scope economies and diversity is recognised then there is not necessarily a need either for large projects or for collaborative projects that cover a large number of countries. Different projects will require very different types of collaboration – for example, collaboration between institutes in two countries or in three countries rather than in six, seven or all the member states of the EU. This suggests that one role for national science policy may be to facilitate international interaction where it is a question of a relatively small number of links. Individual countries may be in the best position to facilitate bilateral or trilateral arrangements rather than the EU which will feel obliged to focus on all the member states of the Union. This suggests that what is needed is a variety of different science policies addressing different aspects or types of international collaboration. Some of these policies will be the responsibility of the EU but others will depend on interaction between individual countries both within and outside of the EU. Such a situation will make any clear definition of roles on a subsidiary principal extremely difficult to specify and, indeed, if a plurality of forms of international institutions and co-operation is allowed to develop, there will inevitably be some overlap and interaction of responsibilities. This suggests that the subsidiarity principle may in this context inhibit the discussion and decisions about the desirable development of both national and EU science policies.

There are further complications in the question of subsidiarity due to the fact that the EU has a wider agenda than the issues discussed so far. It is concerned both to create a European science base with a European community of scientists and also to promote a general integration process and the process of convergence across the EU through encouraging networking that includes the less favoured regions of the EU. There has been some discussion as to whether the aims of cohesion and convergence should be pursued through the Framework Programme when other programmes exist for this purpose (House of Lords 1990). Different member states of the EU have different approaches to the question of the relationship between national and EU science policy. In the UK, policies of attribution and additionality have often been difficult to interpret and have had direct impacts on the nature and distribution of UK science funding. Under the principle of attribution, money that is allocated to EU programmes is then linked back to UK departmental budgets which are at least partially adjusted. This is an interpretation of the role of EU science policy to the effect that, if the EU is undertaking projects because it is more efficient for it to do so, then that is one part of the national science budget and department budgets are accordingly reduced. However, if projects are being undertaken through the EU that would not have been possible or would not have been carried out in an individual country then the decision to cut the UK budget is distorting or at least changing the distribution of UK research projects. Furthermore, if other member states are not adopting such an attribution system – and it appears that only the Netherlands is (House of Commons 1994) – then the balance of the UK science effort relative to that of other countries is being changed due both to EU science policies and to the different reactions of the UK and other member states to those policies.

UK departmental budgets are not cut by the full amount of the UK's contribution to the EU budget, but the procedure involved in additionality is seen as unclear: 'there is no general understanding of how the system operates – or even recognition that it operates at all' (House of Lords 1990). The Treasury has suggested that in the case of R&D, 30-35 pence in the pound of EU funding is truly additional, but it is not clear whether this figure is correct or why it has been chosen (House of Lords 1990). The operation of the attribution and additionality system also suggests a closer relationship between national and EU science policies, priorities and decision-making than is in fact the case. It is quite possible that EU priorities may differ from national priorities but, due to the adjustments in funding already described, EU priorities will take precedence. Furthermore, where there is a major emphasis on collaboration, both for firms and universities, consideration does need to be given to the dangers of insufficient diversity.

Collaboration

In terms of national policy, the UK government has a mixed view of the benefits and desirability of collaboration and the role for government policy in encouraging this. While the UK did place some emphasis on collaboration during the 1980s, in particular through collaborative programmes such as the Alvey programme, considerably less emphasis has been placed on collaboration in recent years. In effect, government policy is seen as one of encouraging industry-university collaboration but not of particularly promoting either collaboration within the scientific community or industry-industry collaboration. The UK Department of Trade and Industry has shifted its innovation support away from industrial collaborative research and new technology towards exploitation, technology transfer and innovation promotion (House of Commons 1993). Only in exceptional cases will support be given to industry-industry collaborations: these are primarily seen as industries' own responsibility. The main DTI support for development of new technologies occurs through industry-science based collaboration and not through any other route. The recent UK Science Policy White Paper (1993) focuses on collaboration either as international collaboration where there are scale economies, or as industry-university collaboration, but not as collaboration within the scientific community. Thus, the purpose of collaboration is to link the science base through to industry more effectively; any wider benefits from collaboration are either not considered or are not seen as the responsibility of government. Concerns have been expressed at the outcomes of the focus on industry-university interaction. It is important that inflexible concentration of university research is prevented and that capacity for long-term basic research is preserved (House of Lords 1993b).

The extent to which national scientific collaboration across universities should be deliberately encouraged is open to question. There are cases both in the natural and social sciences where such collaboration has been encouraged, but in general, it may be argued that if scientists wish to collaborate they should be free to choose their partners and should not be confined to their national environment (Leclerc et al 1992). It is unclear that this point is recognised by the UK government, its view of subsidiarity is that where complementary expertise exists within the country then collaboration should be within the country. This fails to recognise the differences not only in quality and expertise that may lead scientists to choose to collaborate across national boarders but also the importance of mutual respect, trust and understanding which individual scientists may develop through informal meetings and conferences. There is no particular reason or expectation that this informal understanding will be greater among scientists within a nation than across countries, given the international nature of science.

Effects of EU science policy

As has been discussed above, the links between national and EU science policy are complex. It is clear that EU science policy has various effects on national science policy not all of which are fully intentional or fully recognised.

The UK IMPACT (1993) study suggested that EU funding, although perceived as largely beneficial by academics, can make the system more volatile as many scientific groups depend on EU research funds in order to continue; even once successfully established through EU funding, groups are not normally underpinned by UK money. This study also found that although some academics would not have carried out the same research at all in the absence of EU funding, this was a very low percentage – about six per cent – while another 12 per cent would have carried out similar research but with different aims. The IMPACT study also found a significant skills effect of participating in EU projects but expressed concern that this only had an influence at junior level and expressed concern at lack of mobility. The study did not look in any detail at the internal operation and management of networks and so was not in a position to comment on the skills, information and knowledge effects that may arise in this context.

EU science policy may change the distribution of funds for science within the UK with knock-on implications for the behaviour of the UK funding bodies. It may change the research areas and focus of researchers in the UK and it will affect their international links. As industry receives substantial amounts of EU funding, EU policy also affects relationships between UK industry and UK universities. It may be that in some cases where industry would previously have funded university research, it now suggests that industry and universities jointly go to the EU for funding. This will potentially change the nature of the research, including encouraging international collaboration, and it will change the distribution of funds not only across areas but also between firms and universities.

Since the UK operates the system of attribution discussed above and since EU science funds are not fully additional, there is more pressure on UK scientists to apply for European funds and to participate in EU networks than for scientists in most other member states. If this participation is seen as positive then the policy outcome is appropriate. However, it does mean that UK scientists will be participating in EU networks in order to get finance and the scientific and networking motivations may be secondary. Furthermore, there may be substantial differences in the group of scientists who are applying for EU funds relative to those scientists, and also those projects, that would have applied for national funding if it had remained available. Given the EU's focus on competitiveness, there is also a question as to whether the effect of EU funds is to create a more applied focus in UK research and whether this is desirable or otherwise. Overall, it is clear that UK scientists are making a different assessment of the costs and benefits of EU participation due to their different national situation from scientists in other member states.

Although EU science funding remains only a small proportion of national budgets, it is apparent that there are important links between EU and national science policies with various implications for the nature of national science policy and national scientific activity. As discussed above, the question of subsidiarity is complex and unclear in the area of science and technology policy due to the international nature of the scientific community and due to the fact that many of the benefits from collaboration arise due to diversity and complementarity and not simply due to economies of scale. The implications of these issues for both EU and national science policies and for the relationship between the two has yet to be properly assessed and taken into consideration.

Conclusion

This chapter has discussed the increasing importance of collaborative research both for firms and for universities and other research institutes. It considered the main motivations for collaboration and the reasons for the increasing level of collaboration since the 1980s. Increased competitive pressures, together with the risks, complexity and increased interdisciplinarity of much research contribute to an understanding of these developments. Networks are seen as potentially a highly flexible organisational form. Science and technology policy has also played a role, in particular through the European Commission's emphasis on collaboration and networks. Finally, the chapter addressed the question of subsidiarity in science policy and argued that there was no clear dividing line between EU and national responsibilities given that international networks depended as much on diversity as on economies of scale.

2 Co-operative Scientific Research Networks: A framework for analysis

As is clear from the literature, there is a range of motivations for collaborating in research – both for companies and for scientists – and a range of experiences of collaboration. Furthermore, there is a very wide range of types of collaborative activity that companies and scientists have been involved in. Government and EU policy act both to stimulate collaborative research and to influence the type and structure of collaborative research that is undertaken, together with the areas in which it is undertaken. Given the variety and increasing number of collaborative scientific activities, it is important to adopt a general analytical framework that may be used to assess a number of different cases. Much in-depth information about collaboration can only be obtained through detailed case studies, but if these case studies are to be assessed there needs to be a common framework of analysis. This chapter sets out such a framework.

Costs and Benefits

In this project, the main focus is on collaboration between scientists in networks, focusing on the scientific community and not the industrial community (though some networks studied did involve industrialists). One approach is to try to identify the general categories of benefits and costs that may potentially be experienced in co-operative, scientific research networks. There are always inevitably costs in establishing and operating a network, so it is important that there are substantial and appropriately assessed benefits. This immediately raises the question of costs and benefits to whom. The costs and benefits may accrue to the individual researchers in the network, they may accrue at the level of the network as a whole, they may be relevant to the particular scientific discipline or field, and they may have implications for the broader industry or economy beyond the network. The main focus in the current project is on benefits to the researchers and to the network but the other potential levels should also be considered.

Those involved in decisions about networks, from scientists to policy-makers, may be expected to be, in effect, making a cost-benefit judgement when they participate in or establish networks. However, there are a variety of reasons why they may not be making the appropriate cost-benefit judgement viewed from

Table 1	Benefits and costs in co-operative scientific research networks

Benefits	Costs
Finance	Administration
Larger scale research	Time – travel, communication
Wider and/or deeper research	Communication problems
Knowledge, skills, expertise	Excessive central direction/control
Faster progress in research	Constraints on research
Lack of duplication	Lack of flexibility in research
Shared inputs	Loss of diversity
Contacts	Reduced competition
Research outputs	Reduction in external contacts
Intensified scientific interaction	Weak partners/improving competitors' abilities
Status	Future finance/collaboration

the point of view of the society and economy as a whole. Scientists may make an inappropriate judgement about costs and benefits, since a key motivation is likely to be the availability of funding. They should compare the benefits of carrying out research in a network to carrying out either the same research in a different structure or carrying out different research. However, if funds are not available for these latter alternatives, they will then base their decision on the availability of funds.

There may be a similar issue at the level of European policy-makers. When considering establishing a European network, it may be the case that it would be more appropriate to have a series of separate or different initiatives at national level. However, EU policy-makers are not in a position to influence or dictate to this national level; they may therefore decide to go ahead with a network at European level even though it is not the best solution. In this sense, there are a number of complex issues here concerned with questions of the second best and concerned with questions of subsidiarity. The current project cannot address most of the second best and subsidiarity issues directly, in the sense of focusing on the decision-making procedures and influences faced by EU policy-makers. However, by analysing the motivations and results of participation in networks, by comparing national and international networks, and by comparing networks across different disciplines, the current study will contribute to the analysis of these difficult but central policy problems.

Table 1 sets out key perceived benefits and costs to participating in collaborative research networks. This is not intended to provide a completely comprehensive list, but to identify the main likely costs and benefits. Some items appear in effect in both columns. Thus, for example, a network may improve interaction with the scientific community outside the network, it may not affect this interaction, or it may decrease it – if scientists co-operate more with fellow members of the network and less with the wider community than previously. In the latter case, this may to some extent be a cost, and a cost not only to people in the network but a cost to people outside the network. This raises a wider issue which is the dangers that may arise due to exclusion when considerable amounts of science are organised in formal networks. The existence of formal networks may have implications for scientists outside the networks and these implications may not be positive. These wider costs are not always taken into account when assessing the values of collaborative research. However, the case study approach of the current project is not able to investigate these issues of exclusion in greater depth – they can only be noted at this point.

Having set out a framework of potential benefits and costs to co-operative research, the key question is then what determines whether and to what extent those benefits and costs are experienced. There will be a variety of factors determining these outcomes, many of which may be specific to the individual network. However, it is apparent from the existing literature and from analysis of the general structure and aims of such networks what some of the main determining features may be. These include the following main categories:

- Management
- Structure of the network including nature of the interaction and communication in the network
- Aims of the network (narrow, broad, clear etc)
- Researchers' motivations in joining the network
- Funding level
- Characteristics of the researchers involved, including research experience and network experience, and the nature and extent of the complementarity of researchers' skills and of the scientific methods they are used to adopting
- Characteristics of the scientific field and of the scientific community in that scientific field.
- Distance

Each of these potential determinants covers a wide number of issues and factors, and each may contribute to a network being successful or unsuccessful. This is not to say that there is only one right or correct form for any of these particular determinants. Management, for example, is clearly important but management of a network may be loose or may be tight and it is not necessarily the case that one form is always preferable to the other. Similarly, the structure of the network and whether there are high or low levels of interaction, whether that interaction is formal or informal, may be important determinants of success but it is precisely only through more detailed case study that further insight can be gained into which approach is appropriate and when.

Economies of scale, economies of scope and externalities

In general, the benefits and costs of networks can be understood as arising from economies and dis-economies of scale and scope, together with positive and negative externalities or spin-offs that go beyond the scope of the network. Networks may be based predominantly on economies of scale or on economies of scope or they may benefit substantially from both. While the theoretical distinction between scale and scope economies is clear, in the case of networks there may be strong interconnections between the two. Scale economies are present when there are cost advantages of operating at a larger scale for the same activity. Scope economies arise where there are cost advantages to carrying out different activities jointly.

Networks may achieve scale economies due to fixed cost advantages of not duplicating equipment, tests and samples. The use of shared inputs acts as a scale economy. Scale economies may also arise where a much larger project is carried out through bringing together a different number of scientists and laboratories – a bigger and/or faster research project may then be undertaken than would have been possible by one laboratory. Where a project is not feasible for an individual laboratory not because of size but due to the variety of skills and knowledge required, then there are scope economies.

In many, if not most, cases, networks depend on and benefit from substantial amounts of interaction. In some cases, however, projects within a network may

Table 2   An analytical framework for co-operative research

Scale economies & dis-economies	Scope economies & dis-economies	Externalities
Finance	Knowledge, skills/expertise	Status
Larger project: wider area; faster progress; lack of duplication	Wider and deeper research	Knowledge relevant to other research
Shared inputs	Intensified interaction	Future funds/collaboration
Contacts	Weak partners' contributions	Effects of research focus on future research/career
Administration		Effects on scientists beyond the network
Communication		Contacts
Travel		Competition effects
Common framework/ top-down control: distortion of research area and lack of flexibility		

Table 3   Co-operative research assessment matrix

Outcomes	Individuals	Network	Scientific area	Industry/ economy
Scale economies and dis-economies				
Scope economies and dis-economies				
Externalities				

be relatively distinct and interaction can be low. In these instances, the justification for a network is that there are benefits from overall co-ordination and from different projects or different institutes working within a common framework. Benefits from co-ordination may go under the scale or scope category, depending on whether the different projects are essentially working in the same area and field or whether they involve different projects or expertise. These distinctions may become less clear in practice. While differences in knowledge and skills form the basis for scope economies, there may also be relevant scale economies. For example, if a network involves pooling of knowledge, so that there is in some sense a pooled knowledge capital stock, then this may act as a fixed cost which all parts of the network can benefit from without having to duplicate. In this case, there are then also associated scale as well as scope economies.

It is possible to re-allocate the costs and benefits set out in table one to the categories of scale, scope and externalities. This is done in Table 2. Finance is included under the scale economy heading. Although this is not a scale economy benefit in terms of the research in the network, it is a benefit to individual scientists in participating in the network and can be said to derive from the fact of network participation, i.e. participation in a larger scale research project than otherwise. Each of the cost-benefit categories is allocated to one of the scale, scope and externalities categories with the proviso that there may be some overlap between these categories.

As discussed above, the question has to be posed as to who is achieving the scale economies or scope economies. It may be at the level of the individual researcher, it may be at the level of the scientific community as a whole. Furthermore, what is an economy at one level may be a dis-economy at another level. For example, scientists within the network may benefit from scale economies, but the overall impact on the scientific area could constitute a negative scale economy or externality, depending on its effects on other scientists or on specialisation. It is possible to envisage a broad matrix where the effects of participating in a network and the different levels that may be affected by the network are set out. Table 3 sets out such a matrix. Ideally, evaluation of co-operative research should include all elements of the matrix. This poses particular challenges at the broader levels, but at all levels there are difficulties both in assessing effects and constructing appropriate counterfactuals. Detailed case studies represent one way to assess the operation and outcomes of co-operative research at the level of the individual network, but broader studies are needed to look at the wider levels beyond this. Overall, collaboration in research is likely to have major implications for the research process, but there has been a lack of analysis of the effects on R&D (Hingel 1993).

Choice of research structure
Both economies of scale and economies of scope can explain why some research is carried out in networks rather than by scientists working separately or in their own groups in individual institutes. However, benefits from collaboration can be achieved through other means, in particular through bringing different groups and people together to undertake research in one site or location or through more informal interaction than occurs in networks, i.e. through the main informal channels of academic communication: conferences, seminar series, scientific societies. The question, therefore, arises as to why and whether networks should be chosen as a form of collaboration relative to the other possibilities that exist. To answer this question, it is helpful to categorise networks into three main types as follows:
(i)   separate but related research projects;
(ii) interdependent projects/joint research;
(iii) separate projects whose output taken together also constitutes a project – and the overall project imposes constraints or conditions on the individual projects.

In the first type of network, (i), the benefits of participating in a network will be largely due to the information and communication benefits of being closely linked to other people working in similar or related topics or areas. In the second type of network, (ii), the benefits clearly arise directly from the possibility of doing joint research. In the third type of network, (iii), there may be expected to be relatively low interaction, but the benefits arise due to the nature of the overall output that comes from undertaking the individual projects in a common framework. Types (i) and (ii) will benefit in particular from scope economies, though scale economies may also be present. Type (iii) may be seen to benefit largely from economies of scale, though if projects are highly differentiated, but without interaction, an economies of scope classification could be used. Networks may be 'hybrids' of these different types. Thus, a network may consist of a mixture of types (ii) and (iii) where some of the projects are interdependent or there are sub- groups that are interdependent but not all groups are interdependent and yet nevertheless they exist in one common network. Similarly, types (i) and (ii) may also exist within one network, where some projects are closely inter-related and others benefit from being in the same network but do not constitute joint research.

None of these three types of interaction necessarily need to take place via networks. Type (i) will be most appropriate as a network if there is a relatively high degree of relatedness between the projects and if more intense information exchange and knowledge transfer occurs within a network. If these conditions are not met, then informal interaction mechanisms such as conferences and societies may have lower costs be equally effective and, therefore, be preferable. Types (ii) and (iii) must either be set up as networks or as research in one site given their closer interaction or common framework. There are various sunk costs associated both with individuals' location and with the establishment of laboratories, and there are also costs of locating researchers together in one appropriate site. Unless the scale economy benefits of locating people in one site are greater than the costs associated with doing that then a network structure will be preferable. Since, in general, researchers are working on a variety of projects and in a variety of networks, the appropriateness and costs of locating researchers together will usually be extremely high and would limit diversity of structures. Networks relative to these two other main forms of collaboration – single site and informal interaction – have the benefit that they allow some scale economy advantages but without loss of diversity and they allow more intense exploitation of diversity than other forms of collaboration.

In their pure forms, types (i) and (ii) depend on diversity and interaction for their main benefits. Type (iii) does not and, apart from individual research benefits in the individual projects, the only benefits to participants are due to the nature of the overall project. In type (iii), the common framework or constraints may narrow diversity. In general, it appears that, the higher the scale economies, the greater the tendency may be to move from a network organisation to a single facility organisation. At the same time, the lower the scope economies, the greater there will be a tendency to move from network organisation to more informal communication channels. Apart from in type (iii) cases, interaction is an essential part of network operation. Understanding the costs and benefits of networks, and analysing their operation, management, communication and results, will depend in part, therefore, on analysing the nature and structure of this interaction.

The case study approach

The current project – focusing on case studies of national and European networks – is concerned not only to assess these networks but to assess the similarities and differences between national and international networks. This analytical framework allows this to be done, since the networks can be compared in terms of which of the general characteristics set out here are most important in their operation and structure.

Given this overall approach to analysing co-operative research, the case study interviews were based on a semi-structured questionnaire with the aim of identifying: the nature and extent of the costs and benefits in each network, which of the scale, scope and externality categories the networks were in, and what the underlying determinants of network operation and outcomes were. The areas covered may be categorised as follows: funding; structure; motivations; nature of the research; effects on the individual research focus and on the wider area; management; communication and interaction; outputs; overall advantages and disadvantages; comparisons between networks and single institute research, and more informal channels of research interaction; and comparisons between national and international networks. These allow the identification of economies of scale, scope and externalities, and of some of their key determinants.

Thus, in carrying out the case studies, basic information on the network was collected in terms of key issues including: the number of institutes involved in the network, the choice of co-ordinator and the co-ordinator's role, the funding sources for the network and the main research aims of the network. The scientists interviewed for the case studies were then posed a series of broad questions within which to discuss their experience of the network. These questions included a discussion of how the network was initially established, how the participants were selected and how the research programme of the network was chosen. Participants were asked about their motivations behind participating in the network. Suggested motivations included: funding in order to cover salaries and equipment, funding for a larger research project, collaboration to enable a larger or different research project to be undertaken than otherwise, collaboration to broaden contacts, collaboration to widen and/or deepen research expertise and any other motivations they wished to put forward. There were questions concerning the research programme of the network, in particular whether scientists' research was the same or similar to research that would have been done anyway, whether it was complementary, new, and of more or less interest than research that would have been done otherwise.

The case studies covered the operation of the network and communication, including: how the network was managed; what the research interdependencies between the institutes were; the formal and informal structures whereby the network met and communicated; and the nature of communication other than through meetings. Participants were asked about any difficulties or problems in communicating and interacting within the network and what underlay those difficulties, including whether problems arose from: geographical spread, differences in research projects, research approaches or other cultural or organisational differences. Co-ordinators were also asked about any particular problems in managing the project and how this differed from managing a project in a single institute. Participants were asked about the main outcomes to date of the collaboration, expected future outcomes and the perceived contributions of different participants. The interviews included a discussion of what the anticipated benefits from the network were and what the actual advantages or disadvantages have been.

There was a discussion with participants of the main differences in doing research in an international and a national network and relative to research in a single institute and what underlay these differences if they did exist – whether it was to do with differences in expertise, communication methods, structure of research project, difficulties in network operation or any other motivations. Participants were also asked whether the research programme of the network encouraged synergy and a movement towards common understanding of key issues of research and whether that had resulted in a reduction in the diversity of research or research approaches adopted. They were asked about participation in the network in terms of its effects on contact and communication with other researchers in the field and in terms of whether the interaction within the network was different from how scientists would have interacted in the absence of the network. There was also discussion of whether being in a network increased collaboration and reduced competition between scientists.

Overall, this approach allowed a wide-ranging analysis of each of the case study networks to be undertaken. The eight case studies consisted of four pairs of studies – in each pair there was one UK and one European network (in most but not all cases, EU networks). The four pairs covered the following areas: computational aerodynamics, biotechnology, environmental science, and social science (the study of small and medium-sized enterprises). The case studies are reported in the following chapters. The concluding chapter undertakes a comparative analysis of the results of the case studies.

3 Modelling Turbulence (Case Study 1)

This case study is in the area of aerospace. It is an example of a network of university scientists in the UK working through and with a large industrial company – in this case Rolls-Royce. The case study offers a number of interesting insights into collaborative scientific research networks, partly due to the role of the industrial company and partly due to the evolutionary nature of this particular network.

An overview of the network
The network consists of four main groups. Interviews were undertaken with representatives of each group. The key institution is Rolls-Royce. It funds the network, but the various research projects also receive support in some cases from the Science and Engineering Research Council (SERC) and in some cases are also co-funded with the Defence Research Association (DRA). The other three institutions are all British universities. Much of the funding involves PhD students and some post-doctoral funds. The network is informal in the sense that it was not initially set up as a network and it remains formally a set of bilateral contracts between each of the universities and Rolls-Royce. The network as such has been in existence since 1990, but the partners have links to Rolls-Royce going back to beyond the start of this particular network. Due to the informal nature of this network, there is also no clear end date to its work as different contracts may be renewed at different points in the next few years.

The ultimate aim of the network is to improve Rolls-Royce's ability to design components of their jet engines, in particular to design the different blades which are inside the engine – compressor blades and turbine blades. The network is focused on the development of turbulence models especially looking at laminar to turbulence transition models for use within computational fluid dynamics software. The turbine and compressor blades in an engine are transitional over a large part of their surface and it is extremely difficult to predict exactly where that transition point will occur or does occur. The performance of the blade, and ultimately of the engine as a whole, particularly its fuel consumption, is critically dependant on the design of the blade and the location of the transition point. Blades can only be designed with the transition point in the most effective place if it is known where to predict it will be. The latest turbulence models aim to predict the transition points.

In current design, simpler empirical models are used based on previous experience and testing. The aim is to build increasingly powerful computer models to improve their predictiveness and their appropriateness in actual use. The research is benefiting from the increased size and power of computers, and one main result is that the actual numerical errors which are always present in these programmes will get smaller and smaller. Nevertheless there are still limits on these research approaches due to computer size. The research aims have clear application to Rolls-Royce's business, but much of this research is highly complex and fundamental in nature and while there may be a continuous output from the work of relevant information and ideas, much of it is many years away from actual application to existing design procedures.

This network is an example of a company funding basic and strategic research with very clear but long run aims. Rolls-Royce would not have the resources in-house to do a lot of the fairly basic research involved and from a company perspective it would perhaps be seen as too fundamental for in-house work anyway. The three university groups involved in the research contribute different expertise to the overall programme. In particular, one of the groups is concerned with the simulation of turbulence rather than the construction of models of turbulence. These two approaches to turbulence have developed separately and been seen by many scientists as competing. One of the results of this network is interaction between scientists taking a modelling and scientists taking a simulation approach to turbulence, with positive and productive results.

Establishing the network
The network was not set up as a formal network at one particular point in time. Rolls-Royce had already had considerable contact with the three universities involved and with the key scientists leading the work at the three universities. One university, in particular, was seen as the leading centre for turbulence and Rolls-Royce also funds a lectureship at this university. In another of the universities, Rolls-Royce had had contact and projects with the key scientist over a period of about ten years. In the third university – the simulation group – the length of contact had been shorter, but at least one previous project had been worked on between Rolls-Royce and that group.

While Rolls-Royce established bilateral links with each of the three universities, it was keen to see interaction between the groups both indirectly through Rolls-Royce and directly through the groups interacting with each other and working with each other where appropriate on the research issues. So although the network was not initially conceived as a network as such, the lead company was keen to see that this sort of interaction did develop and sees the informal network that has resulted as an example of a successful network.

It did not appear that such a network could be established in anything other than an informal or evolutionary way. Rolls-Royce saw its links and understanding of the universities' work developing over time in an evolutionary way and that it would not have been possible to establish the network from a particular point in a very formal manner. An evolutionary approach to setting up the network is effective since it means new people can be identified over time through existing contacts and experience and those people can be approached and if appropriate slowly be brought into new networks. The structures they use in this sense are fairly informal and evolutionary. While the objectives are clear and well structured, if they need to be changed Rolls-Royce will move to change them. This contrasts with more formal networks where scientists may come together mainly to obtain funding, and where technical objectives may be less focused and/or difficult to develop or change over time. Rolls-Royce also has wider international collaborative links in Europe and the rest of the world, in which some members of the UK network play a part.

Motivations
Motivations for participating in the network varied between Rolls-Royce and the universities. Rolls-Royce aimed to widen its research base and access more specialist and basic research expertise than it has in-house. Rolls-Royce benefits considerably from the contacts with the universities. It funds a number of PhD students and gets benefits potentially from later employing some of these people and from the contact with the various people who are supervising the students. In addition, those contacts give them contacts elsewhere and ensure that they have a flow of ideas coming in. The majority of the basic software that they use has come from the universities. In general, they are not giving technology to the universities, though they are keen for them to work on the models that they have and codes that they have in-house. Rolls-Royce benefits from the fact that the university people have much wider scientific networks, through conferences and other meetings and interactions.

From the university perspective, funding for research is one of the motivations for undertaking such projects. However, in all cases the scientists involved considered that they would have been doing similar research anyway and may well have applied for funding for similar projects elsewhere. The fact that the overall strategic aims of the research were set by an industrial company was not seen as problematic due to the nature of aeronautical engineering research. Indeed, it was seen as welcome and positive to work on the industry's identified future needs. This gave a clear focus to research, at the same time much of the research remained highly complex and fundamental or basic in nature. Working with such a company, was seen as a positive motivation for undertaking such research. Since the network evolved over time, being in a network was not an initial motivation for undertaking the research, but was perceived as a positive experience as the network developed. Nevertheless, some of the scientists involved could see that Rolls-Royce's strategy in terms of the transition and turbulence work was involving other groups who were key players in that area in the UK and in this sense this was a motivation for participating in this work.

Economies of scope were, therefore, more important than economies of scale in terms of knowledge and information sharing.

Working so closely with an industrial company, and where the company is setting the strategic aims means that scientists are, in some sense, not in control of the overall research agenda. It was seen, however, as important to be flexible in this regard. Too narrow a focus on one area, if it did not remain the area of interest to industry, could in the end lead to scientific work being in some sense redundant. One way to avoid such problems is to be aware not only of scientific developments in the field but also of industry's changing needs and demands over time.

Management and communication

Rolls-Royce provided the co-ordination for this network. The network can be described as being at the same time both tightly and loosely managed. It is tightly managed, in that Rolls-Royce's overall strategic aims are very clear and the specific research projects with each partner are also clearly set out. There are regular meetings with the university groups and communication between the meetings as well. The co-ordinator at Rolls-Royce sets up so-called research brochures with each university which defines what work is to be done, what is funded, the time scales and so forth. At the same time, the management structure can be said to be loose in the sense that once a project has been established and the aims set out, there is not day-to-day interaction with Rolls-Royce nor are there completely formal, inflexible structures in terms of reporting and meetings. Rolls-Royce was seen by the scientists as being relatively unusual for an industrial company, or at least for a British industrial company, in having both a very clear corporate technology strategy, and also a clear idea of the nature of the fundamental research scientists were engaged in and the ways in which university scientists work. Where other firms are seen as maybe not having sufficient PhDs and university experience to understand science and universities, Rolls-Royce was seen as being particularly good in this regard. This meant that while Rolls-Royce did expect feed-back through fairly regular meetings and reports, it was not running the network to the sort of tight structures that would be experienced within a firm, including within Rolls-Royce itself. It was also in a position to understand when research timing took longer than expected. However, it was well understood by the scientists that while the company may be understanding of inevitable delays at times in research, they would overall expect delivery of results within the time-frame of contracts, especially if researchers were to maintain a positive relationship with the company in the future.

The network as a whole does not have any formal meetings. Rolls-Royce has review meetings with each of the universities about every three months. It is common for at least one of the other universities to be present at the meeting with another university. Two of the university groups are developing advanced turbulence models for transition but in rather different ways. It is not clear what is the best approach – one is standard and one is novel and the two are to be compared in great detail – this leads to a lot of communication and interaction.

In addition to these meetings, there is considerable communication via telephone, fax and e-mail. Furthermore, one particular member of the network acts as a key linchpin in the network, frequently travelling to the other two universities. At the instigation of Rolls-Royce, this scientist was also awarded a visiting fellowship at one of the other universities to further intensify the interaction between the two. Rolls-Royce actually encourages communication directly between the university groups and not always through Rolls-Royce – they see in the long run that they may get more out of the network operating in this way. Meetings are not on a strict time basis and they see some of the university groups more often than others depending on the nature of the work and the interaction.

Overall, there did not appear to be any communication problems or any difficulties arising from the fact that scientists were working at three different university sites which were also separate from the location of the Rolls-Royce group. There was somewhat less communication between two of the universities, but even here results were being fed from one team's work into the other team's work.

It was seen in some ways as an advantage to be working in three different universities since the different environments – interacting with other scientists and other projects – meant that different bits of information and different view-points were obtained and could be occasionally fed in. This helps to create variety as well as synergy whereas it is possible, as is observed sometimes in the US, to find a lot of groups all working on a very similar narrow area, just because it was the area being funded, with consequent loss of diversity.

The overall operation of the network was seen as fairly flexible. The small scale of the network in terms of numbers of groups and its being led by one industrial company meant that if the development of the research called for changes in the research programme or in particular projects this was relatively easy to achieve. While there was this flexibility in changing direction, this could also have the potential to cause problems if scientists are being asked to switch the direction of their research. In general, this was not seen as a problem and would be the result of two way discussion between the groups involved and Rolls-Royce. Being in an informal flexible network also involved much less administration and paper work leaving much more time for the actual research. The leadership and management of the network was seen as central to its effective functioning. This network benefited from being small but management and leadership remain important.

Results from the network
The work in the network is on-going and fairly fundamental. Nevertheless, the groups involved have produced various results and developments in modelling so far which have fed both into each others' research and also into research and

models at Rolls-Royce. In terms of outcomes so far, they have a fairly good idea of modelling approaches that are not appropriate, and in addition a very good idea of what the actual physical description limitations are of the modelling that is being used. This means that they have narrowed down from the very wide range of models to two or three key ones and their derivatives which are likely to provide the best route to getting the true predictions for the flows that Rolls are interested in. The scientists in the universities may publish the results of what they are doing, but it is usually written into their contracts that there may have to be a delay of one or two years in publication. Rolls-Royce has a publications committee which will consider scientific articles before they are submitted to journals, but issues of commercial confidentiality rarely arise due to the distance from final application of the research that is being undertaken.

Other developments from the network include the benefit that the scientists involved have obtained from interacting and working with each other. This interaction is seen as substantially more intense than that that would come from informal scientific interaction through conferences, common meetings and workshops. The fact that they are working within a common programme with a clear overall technological aim meant that the amount of assistance and interaction was much more intense than normal. In particular, the development of interaction between a simulation approach to turbulence and the modelling approach to turbulence was seen as a major step forward in the area. Rolls-Royce is perceived as benefiting from and developing their relationships with university groups in a very focused way, more than in many other companies. Rolls-Royce was seen as providing the strategic vision but not as trying to take the scientists over or direct them in a way that would be too close or inappropriate.

Collaboration allowed scientists to cover a much larger area of ground than they could on their own which also results in generation of more papers. Benefits from collaboration, in terms of additional contacts outside the network, depend on where scientists are located. In the case of this network, the scientists were located in leading universities, where access to a wide range of contacts was already possible. In addition, in parallel with the network but beginning somewhat later, one of the members – the scientist acting as a linchpin between the university teams – was involved in establishing a European Special Interest Group, funded by Rolls-Royce. This Special Interest Group provides Rolls-Royce with access to, and contacts with, a wide range of European scientists in the area. It also involves some input from groups in the US, Japan and Australia.

For at least one of the scientists, the network opened up a new area of research and one that was likely to be more important in the future. This was seen as broadening out into a new area from an existing base, not completely changing the direction of the research. This change of area was not seen as problematic; in a scientific sense some of the previous research may have been as interesting but it was seen as important to focus on projects and topics that were timely from the industrys' point of view. There was no general view that there was a danger from collaboration of less diversity in scientific research. However, this was in part because the universities and scientists involved did have funding from other sources and were in a position to follow up other lines of enquiry in some cases if they thought they were of interest. There was also a perception that people now have genuine enthusiasm for collaborating when five or six years ago they did not think about science and research in that way. All participants saw collaboration essentially in a very positive light.

The network had therefore led to increased interaction and collaboration between the three university groups and with the Rolls-Royce research group. This collaboration was not seen in any sense as damaging to the normal competition perceived to exist in the scientific world. Collaboration was seen as an important route to widening knowledge and participating in a broader research project – both scale and scope economies were important – with the main emphasis on the benefits from knowledge exchange i.e. scope economies.

There was a view that, in general, it was more difficult to participate in collaborative research within the UK. The European Union situation was clearly well focused on collaboration, whereas while the SERC was seen to encourage meetings and networks for communication, it was less clear whether it would fund many collaborative projects across a number of universities. However, the situation was perceived to be worse in the United States where levels of competition for funds for individual projects were perceived as possibly damaging leading not only to cut-throat competition but also potentially to excessive focus on one narrow area of the field.

One problem that exists in this area is the effects of the recession, which have been particularly severe in the aerospace sector. This may affect industrial funding of basic research in the universities. Science policy institutions need to be alert to these pressures.

Conclusions

This case study is an example of a successful scientific research network established and co-ordinated by a large industrial company. Its success derives in part from the evolutionary and flexible nature of the network. This was an advantage both to the university scientists and to the company itself. The main advantages from the company's side lie in access to fundamental scientific expertise and a flow of new ideas and work. There are also some advantages in access to newly qualified scientists. The university scientists benefit both from the increased interaction and knowledge transfer between the different groups and from working for and with the technology strategy of a key company in the area. The main benefits derive from economies of scope.

4 European Computational Aerodynamics Research Programme (Case Study 2)

This case study is of the European Computational Aerodynamics Research Programme (ECARP). In terms of number of participants, the network is one of the largest, if not the largest, for a specific European Commission project, including almost all the EU countries and two EFTA countries. It brings together all the major European airframe manufacturers and a large number of European universities and national research institutes. The case study is of particular interest due to the size factor, due to the large scale interaction between universities and industry throughout the EU, and because of the fact that the industry faced major recessionary problems as the project got under way. Interviews were undertaken with the co-ordinator and with six participating universities, as the focus of the project was on university scientists not research in industry.

An overview of the network
The network is part of the Commission's BRITE-EURAM programme. Within this, there is a special aeronautics area and ECARP comes under this. ECARP grew out of three prior projects in this area. These three previous projects were seen as a pilot phase of an ongoing programme. The overall programme is focused on addressing the issues of the computational simulation of airflows. ECARP is then the second or intermediate phase of this European aeronautics programme. It was intended that there would be a final phase to the programme, but it is unclear as yet whether or not this will take place. There are, inevitably, questions raised as to whether it is appropriate for aeronautics to have its own special programme within the overall BRITE-EURAM area.

The fact that ECARP is an intermediate phase of the aeronautics programme is useful for the purposes of this case study, since it is possible, in part, to compare it to the three projects in the pilot phase from which it grew. The projects in the pilot phase are seen by participants as having been generally very successful. They were smaller than ECARP, having a size of about ten to fifteen participants in each. These three pilot projects looked at the questions of optimal design, validation, and grid generation.

The decision to bring together these three strands of the aeronautics programme into one larger overall project – ECARP – was taken after discussion between industry representatives and the Commission. The main motivation for bringing the three strands together was primarily one of ensuring funding for the overall aeronautics programme. There was a concern that if four or five projects were put in as separate bids they may not all get funded. Since the four or five projects were interdependent, this outcome would have been damaging for the overall rationale and success of the programme as a whole. There was also a view that, by bringing the three main strands together, this might encourage more interaction across the three strands.

The main problems that arose due to this approach were, firstly, that ECARP was then an extremely large network raising a number of managerial and operational difficulties, and, secondly, although ECARP was successful in receiving funding it received only a small part of the funding it had applied for – about ECU 2.27 million. Although it was intended as the intermediate phase of the aeronautics programme, the amount of money was roughly the same as in the pilot phase. The advantage to the manufacturers in going for a larger umbrella programme for ECARP was that it gave more control back to the manufacturers. Even if the Commission – as happened – rejected the size of the overall bid, the manufacturers would then have scope to discuss how to reduce or amend the programme. If, in contrast, four or five separate bids had been submitted, then the Commission could reject one or two of those and it would be impossible for the industry to come back and try to modify its proposals.

There is substantial discussion and communication between the aeronautics industry in Europe and the European Commission when such overall programmes are established. There is a management committee of the aeronautics industry which is made up of senior research director representatives from each of the major European airframe companies. This management committee discusses the overall requirements for aeronautics research, not only those for computational aerodynamics, and then presents its views to the Commission. The Commission will take into account these representations when it sets out its core research programme and the industry management committee will then co-ordinate responses to the Commission's programme to ensure both that industry's requirements are met and that there are not conflicting or duplicating proposals submitted.

Research aims
The main aim of ECARP is to reduce the design times and design costs of aircraft, from the aerodynamics point of view only, by at least a factor of a 50 per cent reduction, with the end result that it will take half the time that it now requires to design an aircraft. It is anticipated that this 50 per cent reduction could be achieved in about five years time. In this research area initially, many years ago, wind tunnels were the main means of research. They remain very expensive. Research moved on to a combination of wind tunnels and simple prediction

methods. There are limits to how much further the simple prediction methods can be developed, and the major emphasis is now to focus on more computationally-intensive tasks and approaches. Using computational aerodynamics, the industry has achieved fairly substantial savings in the past ten to fifteen years, but there is perceived to be substantial scope for further progress in this area. There has already been substantial investment in this area, and so much of the research is at the stage of actual implementation within an industrial environment. At the same time, there remain some fairly fundamental research questions to be addressed. The research within the network can, therefore, be described as a mixture of basic/strategic and applied research.

Establishing the network
There are 36 individual organisations from 12 separate countries in ECARP. There are ten EU countries together with Sweden and Norway. The only two EU countries not represented are Luxembourg and Portugal, and Portugal will shortly be brought into the project with a small amount of new money that has been made available. The project is managed and co-ordinated by British Aerospace. In establishing the proposal and the network, there was a project management committee which included representatives of all the airframe manufacturers. Working to this project management committee, was a group of four people including the overall ECARP co-ordinator at British Aerospace – these four people had been responsible for the management of the previous pilot projects. There was an initial meeting with the project management committee to discuss and negotiate an overall structure for a research programme, and it was then for these four people to assemble a more specific programme in line with the directions of the project management committee. This was a time-consuming and difficult process since it involved negotiations and discussions with different companies and people who have their own specific national and company interests. This overall process took about nine months.

In terms of selecting universities and research institutes to participate in the network, there was a lot of prior experience to draw on, including not only the three existing pilot projects but also substantial experience of the major airframe manufacturers of working with particular universities and research groups. In establishing ECARP, there was also the intention of selecting centres of research excellence that were involved in more than one of the three main research areas in order to encourage the interdependent and interdisciplinary elements of the project. This aim arose out of experience from the pilot phase where institutes had worked very effectively in their own area but there was a lack of communication across the three projects.

Experience in establishing and managing networks was seen as essential in setting up a major network of this kind, including having experience of, and realistic expectations about, working with and negotiating with other partners and with the Commission itself. At overall corporate level, co-ordination of the network is seen as a plus for a company, but at the research manager level the complexity and difficulty of the management task means that having the role of co-ordinator is not always seen as desirable. In the case of ECARP, the original co-ordinator was a representative of regional aircraft within British Aerospace. Due to this part of British Aerospace being merged with a Taiwanese company, the co-ordinator's position had to be moved and so it was moved within British Aerospace to another research centre that was also involved in the programme.

Motivations

In such a large network, motivations for participation will inevitably vary. Since it is so large and comprehensive a network, one motivation is to ensure that a research group or institute is part of such a key grouping of leading firms and universities in the aeronautics area. Another reason for collaboration is because there are scale economies in this research and the capital requirements are high. Funding is a major consideration in many cases – the universities get full 100 per cent funding and industry gets 50 per cent funding. It was suggested – by a university scientist – that industry may, in effect, be getting more than 50 per cent funding as the Commission allows industry to charge substantially more for research costs than universities. On the other hand, it was also suggested that, in the US, 100 per cent funding for industry was provided and that this resulted in a different sort of commitment to the network from the companies involved – they were less focused on trying to ensure they got a return to their own company from their investment and more committed to the network as a whole.

Not all groups involved in the network participated to obtain funding. As ECARP did not receive all the funds it had applied for, many groups who had participated fully in the pilot phase received enough funds in the intermediate phase only to participate in meetings and for travel and communication. Others received very small amounts of funding, so that they could only continue their research by also drawing on resources from other areas. These groups continued to participate in part to remain within the wider network both for information reasons and in the hope of future funding.

Other motivations for participating in the network were to benefit from the variety of expertise and approaches of the different participants, access to up-to-date R&D information and reduction of the risks involved in complex R&D. From the universities' point of view, working with industry was generally seen as welcome and as a way of focusing basic and strategic research on the areas where industry needed to develop. Working with industry is beneficial because information is obtained about the problems that they are encountering and focusing on. It forces the research world to be more focused and more goal oriented. Also the problems of industry are usually very complex and they have requirements that may be generally far beyond what is currently feasible. From the industry point of view, the network enables companies to work together at a technical level, but it also enables insights to be obtained into the way different organisations operate and function, which is, in fact, often quite valuable information. Furthermore, it allows access to a wide range of university expertise.

Status from participating in European projects can be important but this status is not dependent on being in one particular project. University scientists are usually involved in a number of networks.

Motivations, therefore, include anticipated benefits from economies of scale and scope, with some externalities in terms of future funding and importance of establishing/maintaining links within the network.

Research

The participants interviewed mostly considered that they would have been likely to do similar research in the absence of the network. However, this was not true in all cases – in at least one case it had allowed development into a new area, and in other cases funding may have been more difficult or slower to obtain. Furthermore, some of the research could only have been done through a network. For example, it had been possible in the pilot phase to ensure that different research groups all worked with the same grid measurements and specifications making work directly comparable which normally would not be possible. This was seen as a major benefit and advance in the area. This could only happen in a tightly co-ordinated network. One participant thought research results may be accepted more widely if they were produced from a broad base of research institutions.

There were some views that the research programme of ECARP could be better focused – this was partly a problem due to size. At least one participant saw the participation of some countries as being unproductive scientifically since they were behind in the field. There were also some differences between the needs and interests of the regional airframe manufacturers relative to those of the airbus consortium. In the absence of the Commission programme, this range of industry and university collaboration would not have been set up for this particular piece of research.

Management and communication

The management of such a large network is an extremely difficult and challenging task. Even with a de-centralised structure of management, there are many problems of co-ordination and communication and more than one participant expressed the view that networks of a size greater than about 15 participants were possibly or probably undesirable.

ECARP has seven main areas. The three main areas are those from the three previous pilot projects – mesh generation, validation and optimum design. In addition there are two so- called concerted actions which are essentially aimed at allowing people to come together simply to meet and discuss, and there are two support activities, one of which is super-computing support. In addition to the overall project co-ordinator, there are co-ordinators for each of the three main areas and they have sub-co-ordinators working below them. The co-ordinator's role is, in a sense, as an arbitrator. His role is also as a first point of contact with the Commission. It is important for the co-ordinator to have a very project-focused set of aims and objectives, and not to let his firm's aims dominate. There are a number of formal meetings with the Commission. There is a project management committee that includes representatives of the main firms. The co-ordinator has an informal discussion with the Commission representative prior to meetings. The committee meets every six months with occasional additional meetings.

The ECARP project runs for two years. This was generally seen as too short a time period, even in smaller networks there is substantial learning to be gone through before a network is operating in the most effective and valuable way possible and this is even more true in the case of a larger network. The network as a whole meets formally every six months for about two days. Participants have to supply a quarterly report every three months which gives a short account of what they have achieved in the previous three months. This is submitted to the co-ordinators who then pass it on to the overall project manager. For the six-monthly meetings, longer reports have to be submitted and these are distributed to participants at the six-monthly meetings. These six-monthly meetings are inevitably difficult to set up and to manage. It is very difficult for a large number of different groups to make presentations within a short period of time on a wide range of areas. Nevertheless, participants mostly found these meetings useful as a way of being informed of what was going on within the network and as a means of interacting with people within the larger network.

Outside of these formal six-monthly meetings, different groups interacted to different extents with other groups. Research groups tended to interact in smaller working groups within one of the main areas. This might mean that any one group is interacting most intensively with anywhere from about two to five other groups. These groups may meet informally and they will communicate through telephone, fax and e-mail. Overall this more intensive interaction between a very small number of groups suggests that there are limited benefits overall to being in such a large network – the economies of scale and scope occur at levels substantially below the overall size of the network.

There was only limited communication by e-mail. This was largely due to the fact that industry will not use e-mail, partly as it remains concerned about confidentiality issues. Universities use e-mail much more readily. Especially in such a large network, e-mail has the potential to facilitate enormously the management task. Even with the use of fax, sending out faxes to 36 organisations is a major task, whereas comparable e-mail communication is clearly much easier.

There were, in general, no apparent major problems of communication between universities and industry. Time was needed initially – in the first phase – to reach agreement on what cases to consider and other questions about how the work should be done. One common issue was that universities may expect and be used to working on different time-scales to industry. It was felt that both sides needed to recognise the different approaches in university and industry on such questions and that learning this takes time. The extent to which industry was perceived as understanding the university research environment, including having staff with experience in universities at a post-doctoral level, was seen to vary by countries. In making on-going decisions about undertaking particular pieces of work or test cases, it was thought that universities could respond more flexibly than industry.

It was seen as very important who is representing the firms – whether representatives are senior managers who can take decisions. There was no general concern that working with industry meant that universities' research would be too narrowly focused in the short term, though this was recognised as a potential risk to which participants should remain alert. European programmes are seen as increasingly focused towards industry and the danger will be if research does get too short term. Computational fluid dynamics were developed 95% in universities – industry was not the driving force – and this is an important example of keeping the research focus medium to long-term. Academics were seen as likely to explore a wider, more diverse range of research paths. There was also a potential risk that short term pressures could come from the Commission itself, especially if it is keen to see concrete results relevant to economic performance from its research project. Some university scientists thought there was inevitably a balance between trying to do what industry wants and trying to develop a coherent, scientifically valid, research programme.

It was also thought that universities varied in their understanding of industry and how industry operated. As research funding becomes tighter, some universities participate in networks solely in order to obtain funding and are not necessarily very receptive to the demands of interaction with industry. At the same time, as the aeronautics industry hit severe problems in the recession, they were seen as less willing or able than previously to fund universities' research and were more keen to encourage universities to obtain funding from the EU. European schemes can, in this way, change the wider university-industry relationship since industry may then ask universities to go to the EU for funds. Also if, as in the UK, funds are not always fully additional, universities may get even less funding because industry gets some of the money previously available for universities.

The severity of the recession and the difficulties facing the aerospace industry had major implications for the network. The aircraft industry in Europe was said to be in 'a catastrophic state'. Companies had to cope with considerable pressures on their own funding and their wider research funding. In some of the main partners, major cuts were taking place in R&D funding and research groups were being cut back. These pressures made it difficult to give the same level of attention and resources to the network. They also meant that companies participating in the network were much more likely to be fighting to keep whatever resources they could for their company. Under severe market pressure, firms may attempt to narrow the interpretation of their commitment to the network. When additional funds became available, firms were possibly less willing to look at the broader aims of the network in allocating this money, for example to universities, rather they would aim to obtain the money for their own company.

Participants did not see major communication problems in the network. However, one factor impeding communication was the fact that universities tend to be involved in a number of different networks at any one point in time, consequently limited attention may be given to any one of those particular networks. This can lead to delays and inadequate levels of response in communication. In terms of the overall management of the project, communication is inevitably more problematic. The English language as such is not perceived as a problem. However, if explicit instructions on network requirements and reporting requirements are sent out to 36 different institutions, the chances are unfortunately high that one or two people will misunderstand or misinterpret some part of the instructions. Even trying to ensure common reporting formats can be highly problematic and strict, and specific forms have to be used. Understanding the necessity of this can be problematic, particularly sometimes for university administrators. Furthermore, the need to follow common formats and to respond to all information requirements can increase the administrative load on participants quite significantly, and, especially where participants have relatively low funding, this can seem relatively burdensome and time-consuming. Getting reports and responses to a request for information in on time can also be a problem, again, simply due to size. Ensuring that all the benefits that are potentially available in terms of interaction and interdependence are actually obtained is difficult.

Geographical distance in Europe was not seen as a problem but distance is still important in communicating with Japanese and American colleagues. For US and Japanese collaboration, funding would need to be for actually paying people to move to those countries for a period. Mobility between research institutes was seen as feasible for post-docs rather than for higher levels. Even then there may be focal points for research in Europe, places that are more attractive in conditions and science level to researchers. This may be a problem for laboratories on the periphery geographically.

Outputs and interaction

Some groups had made substantial steps forward in their scientific research within ECARP. They had already produced and/or anticipated various papers and reports. Overall, it was not expected that all anticipated outcomes would be achieved but that most of them, possibly about 90 per cent, would be achieved. Participants had benefited from the interaction in the network. Industry was seen in some ways to benefit more than universities from such networks in providing a wider range of contacts as university scientists already had networks through standard informal academic interaction and conferences. Nevertheless, such a network was seen as important to academics as a way of keeping abreast of scientific developments; it was a more intense and effective way of keeping in touch with these developments than attending conferences or obtaining working papers. Conferences were generally broader and consequently networks were seen as the most effective way of knowing about key work in specialised areas. Given the increasing amount of scientific output and the range of the field, it was important to work together with other institutes in order to keep up to date and to share expertise. This represents both an important scope economy and a positive externality.

It was considered that scientists interacted more strongly within a network than they would through the informal scientific networks and would be more willing to provide help, information, data exchange and other joint working than outside of a network. Much of the knowledge and information exchange may be intangible, difficult to quantify, and apparently minor, but overall be of considerable importance. Furthermore, although academic scientists already had their international networks, a network such as ECARP still resulted in them being in touch with new groups or in finding more out about the work of existing groups. The network provided the opportunity to meet people to do other research with and to construct further research projects. It was also seen as an important experience for young scientists in interacting with and learning about the experiences and approaches of other young scientists as well as of other research institutes.

One great advantage of a European network is that it tends to be far richer and more diverse in approach – not necessarily in new ideas but often in different ways of looking at the same problem. In a single institute, people may tend to look at the same problem in the same way. From a managerial point of view, a single institute may be easier because people speak the same organisational language with common goals. In a network, even though people may form a consensus, they will still challenge that consensus in a way that does not occur in individual organisations, and they will look in different ways at that consensus. Diversity represents, therefore, one important basis for economies of scope, and this is an area of advantage for the EU relative to the US and Japan.

It is inevitable that, in such industry-academic networks, substantial amounts of information remain confidential to individual firms. The fact that firms are collaborating in no way undermines their sense that they are competitors. It is well understood by industry and by the Commission that industry will not share information that gives it strategic advantage. This is also the case to some extent with scientists in universities. They are also competitive and while they collaborate more intensely in networks will not provide the specific detailed information that is usually the base of their particular advantage in a specific area. It was also suggested that collaboration at EU level can have the effect of motivating the interaction and overall collective aims more than bilateral collaborations where the justification is only at company level. At European level, there is a different profile and emphasis, and this can spur people on to demonstrate their collaboration.

In general, there was a view that the pilot projects had been more successful as the interaction had been in smaller groups which were in general better funded and that the intermediate phase by comparison was not as successful. There had also been about a 9 month delay between the pilot phase and the intermediate phase. For university groups it can be extremely problematic to keep together a research group, and some had to start from scratch again in order to participate in the intermediate phase – there are clearly serious diseconomies here. The university scientists interviewed see the three strands as still very much separate though there are some cross links. Not all participants considered that the potential benefits from interaction had been achieved and that in this sense the value added from the network was less than it could have been. Overall, a network should be aiming to achieve more than the individual members could achieve on their own – the extent to which this was happening was questioned by some. Making sure the interdependence and interaction really worked was seen as much more difficult than making sure specific technical and administrative functions operated appropriately. There was a perception that for the EC it is enough to show that they have brought European scientists and industry together and that the actual scientific outcomes were not so important. There was also a view that much of the scientific output is not recorded and that there should be more co-ordinated data bases set up recording information and results from the research that is done.

Conclusions

ECARP is one of the largest EU networks in terms of number of participating institutes. Bringing together all the key firms and university groups into a wide network together with other firms and universities that can benefit from and contribute to this core can be seen as a substantial achievement. At the same time, it also has to be seen in part as a large-scale experiment which has demonstrated difficulties with such an ambitious grouping. The large absolute size of the network appears to have been seriously problematic. Management ability is critical in networks both small and large, but it appears that at this sort of scale there are simply too many organisational and communication difficulties to make it an ideal size for a network. Diseconomies of scale are high, and some potential positive scale and scope benefits are not achieved due to the overall size.

Difficulties had also arisen due to the funding for the network being substantially less than that requested, and due to the Commission being keen none the less that the network should be a large network covering the different countries and areas. Networks can suffer operationally and scientifically from these mixture of motivations on the part of the Commission, particularly the attempt to be extremely comprehensive in a geographical and political sense. This, however, is not only a question for the Commission, since it was also seen as potentially important in retaining national government support for the special aeronautics programme that all countries had an involvement and an interest in the programme. Low levels of funding also meant that different groups were participating on different bases in the network and therefore could contribute and benefit from the network to different extents.

In addition to the problems due to size and low funding, further difficulties were introduced by the severity of the recession in the aeronautics area. This posed many problems for the companies participating in the network, and these problems inevitably have had an influence on the operation of the network. The perceived more successful nature of the pilot projects suggests strongly that smaller networks, with much clearer overall aims and focus and therefore much stronger synergies and interdependencies between participating groups, can be much more effective as a network. Nevertheless, the network has produced and will produce substantial research outputs. Participants have benefited from interdependencies, even if not to the extent anticipated. It is clear, however, that the benefits could have been greater and the costs less within a different – smaller – structure.

5   The Animal Cell Biotechnology Club (Case Study 3)

This case study is of the Animal Cell Biotechnology Club. This Club was set up under the initiative of the Biotechnology Directorate of the SERC. Funding came from the SERC and also from participating companies. This case study is of particular interest as it represents an attempt to stimulate and to direct science in the UK in an area seen to be of importance to industry and firms.

Overview
The Animal Cell Biotechnology Club had two phases. The first phase ran from 1987-1990 and involved six companies and five universities. The second phase ran from 1991-1994 and involved three universities. Two of the three universities in the second phase had also participated in the first phase. There was also one further university whose project overlapped the end of the first phase of the Club and the start of the second phase. In the first phase, funding was provided on a 50:50 basis by the participating companies and the SERC. In the second phase the SERC provided two-thirds of the funding. Two of the companies from the first phase did not participate in the second phase and one new company participated in the second phase but was not in the first phase of the Club. The second phase was not yet completed when interviews were undertaken – 1994. Interviews were undertaken with participating universities from both the first and the second phase. Interviews were also undertaken with those involved in the management/steering group side of the Club.

The main aim in initiating the Club was to create a critical mass of university scientists working in this area. This would then provide an important research base that biotechnology companies could form links with, and from which biotechnology companies could obtain high quality, appropriately trained employees. The Club was thus a deliberate attempt to create a mass of expertise in the UK where such expertise had not previously existed. This was seen as important since a large proportion of biotechnology products were coming from animal cell products. The subject is interdisciplinary and involves bringing together various disciplines including biochemistry and microbiology. The general aim was to undertake fairly basic research in the biochemistry and physiology of animal cells. This was not seen necessarily as simply a question of the UK catching up with American and Japanese scientists. In the USA and Japan, much of the research was being undertaken in companies and there were not many university departments in the area. It was, nevertheless, seen as a positive idea to try to establish animal cell biotechnology expertise in the UK university sector even though, initially, it was the companies that actually had the greater expertise. The philosophy of building up a community of experts in the field was seen as being more generally part of the approach of the Biotechnology Directorate of the SERC. Benefits to the companies from participating in the Club were likely to be from the trained staff that would become available and in the medium run from the scientific base that would develop in the universities.

During the selection process one aim was to get both more experienced and less experienced university departments into the field. It was seen as inevitable that there would be a long learning period for scientists coming into the area. It is unclear how much influence the companies had in the selection process which was carried out by the steering group, but some companies already had relationships with particular university departments and this may have had an influence.

It appeared that there may have been difficulties in persuading companies to continue to participate in the second phase of the Club. For some companies, including those that participated only in phase one, there were problems due to recessionary and financial pressures. Companies, were also changing their area of focus and specialisation over time. It was also possible that companies were a little disappointed with the output from phase one and so were more reluctant to participate in the second phase. This case study focused on the scientists, however, and no representatives from participating companies were interviewed.

Motivations

Funding was a major motivation in participating in the Club. All the scientists interviewed, however, also had genuine interests in developing their expertise in this area and saw the Club as providing an interesting and positive way to do this, especially given the interdisciplinary nature of the science and the fact that for many of them this was moving into a relatively new area. The excitement of the science was mentioned by one scientist as a reason for entering the Club. Moving into a new area was also seen as exciting and stimulating. Scientists also saw it as potentially beneficial to be able to work with a number of companies in this area, to form better relationships with them and to benefit from their knowledge and expertise and to find out what the particular interests of the companies were. For a number of scientists there were steep learning curves in the early phases of the Club. Given the aims of the Club, it was the case that scientists' research was going in directions that it may not have done otherwise. This was precisely the underlying aim of the Club. At the same time, at least some scientists said that they might have attempted to gain funds from other sources to undertake some work in this area. Some of the equipment involved in the research in this area is very expensive and so the funding was important in enabling the research to go forward. Clubs were also seen as a positive way forward for research both because they encourage collaboration and because they brought companies together contributing to costs on the basis of common interests. This could then be fed through into a number of universities and not only into one university. In this way they could almost be like mini-societies.

Management and communication

The Club had a steering group on which the companies who had contributed to the Club were represented. Over time, during the first phase, company representatives were changed on about three occasions moving each time to more junior representatives. In addition there were two or three academics on the steering group in the two phases of the Club. In the first phase of the Club, there was a programme co-ordinator who was responsible both for the overall administrative co-ordination of the Club and for the scientific co-ordination between the various participants. In the second phase this role was split, so that there was an administrative co-ordinator and a scientific consultant in addition. The scientific consultant in the second phase was in fact working with a large number of animal cell biotechnology projects beyond the scope of the Club but including the three within the Club.

The role of the steering group appears to have caused some problems in the operation of the Club – in particular, the relationships between the steering group and the participating scientists. The role of the steering group was intended to be primarily a monitoring and advice-giving function rather than strongly directional. Nevertheless, especially in the second phase of the Club the view was expressed that the steering group became much more directional and that this was inappropriate and intrusive. It was considered that there were some differences of views and expectations among the companies on the steering group about what the Club was for – whether it was acting for training purposes and to stimulate basic blue sky research or whether it was going to produce results on much more specific areas of interest to the companies.

The participating universities had to provide six-monthly reports on their research. They also had to participate in six-monthly meetings where they presented an oral report to the steering group. At these six-monthly meetings, different formats were tried. Mostly the universities would make their presentations in front of the other university groups. In earlier stages, the steering group would then respond and make comments. However, the dominant format was one where the steering group would meet separately after receiving the reports, agree their comments, and then deliver these to the universities through the project co-ordinator. In addition, in the first phase of the project there was an annual meeting where a wider number of scientists involved in the projects all came together to meet and discuss their work. These meetings were felt to be very valuable and resulted in the establishment of a British branch of the European Society for Animal Cell Technology. In the second phase there were no annual meetings, but this British section of the European society was by then meeting on an annual basis.

In addition to the six monthly meetings the universities were communicating on an informal basis. This would sometimes involve visits to each others laboratories but more frequently it would involve telephone, fax and e-mail communication. Links and communication between the groups were also encouraged by the programme co-ordinator who would visit the different university sites.

Some disappointment was expressed that there had not been more intensive collaboration and interaction between the different universities involved. In particular, especially at the start of the project, some universities were more up to speed in the area than others and it was felt that there could have been more interaction to the benefit of most participants. The communication and interaction that did exist in both phases was largely of a relatively low scale technical nature rather than anything resembling joint research or joint projects. In the second phase, the three projects were largely independent and so it was seen as inevitable that communication was limited to a certain amount of technical information and sample exchange. Scientists did consider that this communication had been useful and that it was more intense than would have occurred in the absence of the Club. It was unclear how much more could be done to manage or direct such collaboration where universities did not choose to initiate it themselves.

The steering group played a relatively strong role in commenting on, criticising and to some extent directing the scientists' research. Its role in directing research was seen to have grown stronger in the second phase of the programme. It was also considered that both toward the end of the first phase and in the second phase that the steering group had become considerably more critical and that while some of these criticisms were constructive, many were negative. There did appear to be a problem of the nature of the relationships between a number of the universities and the steering group – the relationship appeared to have become fairly antagonistic with deleterious effects on communication, trust and responsiveness of both sides to each other's points of view. It was suggested that there was some division in the steering group between those who understood the need to leave university scientists relatively free and those who wanted to impose a rigid programme of work. In industry if work is being performed badly the response is to manage it more tightly; this may be inappropriate in an academic environment. There was a view that the short-term reporting structure and possibly the short-run perspectives of the industrialists did lead to changes in the way the research was done. There was pressure to produce the results in time for the next meeting. The research was not industrial project work and yet it was being done on an industrial time-scale. The view was also expressed that junior company representatives might be more critical as they are nearer the cutting edge in science and need to impress their line managers.

At the end of the first phase of the Club, three external scientists were brought in to assess the scientific output and work of the participants. These reports on the whole were highly critical. The view was expressed that the science was not always of high quality or innovative. Nor had all universities fulfilled most of the objectives they had set out to fulfil. Some in the Club considered that these criticisms had been too strong, that the external assessors failed to appreciate the overall aims of the Club and that there had, inevitably, been a relatively long learning phase as scientists established themselves in the area. Indeed, one of the projects that received most praise in the assessment, was one that had actually been strongly criticised by the steering group – but since it had limited objectives it appeared to have achieved them. Nevertheless, it is clear that there were problems in the nature and organisation of the science of some of the university groups.

One major effect of these critical reports together with acknowledged problems in some of the phase one projects appears to have been that the steering group then operated more tightly and in a more controlled way in the second phase of the Club. Unfortunately, this led to a deterioration in communication within the Club in phase two, and to a rather negative atmosphere. There was also a view that in the second phase and especially in the second half of the second phase that companies were more concerned to try and obtain results of specific value to them from the Club rather than being committed to basic research. With the SERC reorganisation, there have also been some problems in momentum in the last phase of the Club. The review meeting which should lead into the last year of the Club had been delayed by a number of months at the time of the case study. There was also a view that there had been a loss of continuity between the two phases and so a loss of learning benefits. It was thought that it might have been more productive to keep more of the initial researchers in the first phase in the second phase; to some extent the two phases were like two separate events. The role of the programme co-ordinator in the first phase – responsible for administrative and scientific co-ordination – was seen as important, and the lack of this role in the second phase was deleterious. At the same time, although the problems appear to have been more severe in the second phase, there was also a view that the science had actually been more effective and more 'professional' in the second phase.

The role of the companies in the Club was seen to be both positive and negative. The problems with company participation were perceived to be that because there were a number of companies on the steering group, companies were not providing all the information that they had and that they could, and indeed they did provide more information on a one-to-one basis with university groups when they were on different contracts. Some scientists felt that they were being left to rediscover the wheel, when companies could have told them the results earlier. These issues of company confidentiality in front of other companies are to some extent inevitable but do seem to have been problematic in this case.

The problems with company communication also related to the cell lines that the scientists were working on. There was supposed to be one common cell line across the Club but due to problems with the initial cell line there were at least two. It was considered that companies could have been more helpful in providing cell lines or in providing advice on cell lines. There also appeared to be differences between what companies were prepared to say formally in the steering group meetings and what they would say informally to individual scientists. Especially in the second phase of the project, companies were expressing substantial scepticism informally both about the value of the particular cell line and about the work being undertaken – they saw the main benefits as being in the trained staff that were being produced. One scientist commented on the lack of focused industrial participation – he said that he would have been delighted to have been asked to solve a specific problem. One view was that the Club would have operated more effectively if companies had taken a broader industry perspective, rather than being concerned with their own company interests. Another problem was that commercially the companies were seen to be worried about commercial threats from academics to their patents if the academics were successful in an area directly competitive with the companies. This raises a potentially important and more general point about the role of firms in directing or influencing university research. Serious conflicts of interest may exist which could be detrimental to research in the universities.

There seemed to be a number of causes of the problems in communication within the Club and the negative atmosphere that in many cases resulted. One problem is that the remit of the steering group does not appear to have been understood in the same way by all participants, and the operation and remit as interpreted by the steering group appeared to change over time. Part of the problem here may be to do with varying company objectives, both across companies but also within the same company over time, especially as recessionary and financial pressures grow. The changes in company representatives to more junior representatives may not have helped. Dynamics of a group also develop over time – more than one scientist commented that often criticism seemed to be being made purely for the sake of being seen to be critical. Whether true or not, this perception gives a strong indication of the communication problem that existed. Meetings with the steering group were approached with apprehension. Some of the problems appeared also to depend on different personalities. Another factor underlying the communication problem, may be to do with the overall aims of the Club. The Club was aimed at developing a new area of science within British universities. This, inevitably, involved bringing new people into an area and involved these new people taking substantial time to learn and develop within the area. This is more likely to result in problems in the level and quality of the science being done.

Despite these problems, one university commented that it was important for projects to have clear management and administrative structures, with deadline, milestones and so forth. These were seen as valuable lessons for young scientists and it was also valuable for young scientists to learn to withstand and respond to strong criticism. There was a view expressed by more than one that the Club approach was appropriate in the early phases of research but that it had been inappropriate to carry it on for as long as six years – four was possibly sufficient. After that a looser organisation with more freedom for companies and universities was appropriate.

Despite the problems with the steering group, a number of universities had developed productive and positive relationships on a one-to-one basis with companies involved. These, in general, were found to be much more productive and beneficial with much greater information and knowledge flows and a much more positive atmosphere as a result.

Outputs

Most of the scientists considered they had benefited substantially in research terms from participating in the Club. They had produced results which were published in academic journals and so this was seen as an indicator of successful research. For some groups there had been additional results. It enabled at least one university to establish a strong animal cell biotechnology presence working on a number of projects not only those funded by the Club. However, these benefits derived from their specific projects, and whereas benefits from being in the Club were low – few positive network effects had been achieved. There was also a view that there had been the potential to do substantially more in this area both in terms of the collaboration between the groups in the two phases and in terms of the large number of initial project applications that there had been. Nevertheless, there was seen to be much more international collaboration and interaction as a result. Meeting people from different disciplines was seen as valuable and interesting. Better relationships and contacts with people in industry had also been established. There were probably about double the number of scientists working in UK universities in this area than previously.

In general, the Club was seen as having been moderately successful in achieving its aims in that a critical mass of ability in the animal cell field had now been established and that there were substantial high-quality trained people available for industry. On the other hand, some doubts were expressed as to the overall coherence of the broader science policy objectives in this area. Furthermore, the Club had not operated as an effective network. There had been some problems due to the recession so that while trained staff are now available they were not in all cases able to find jobs. One scientist also commented that although he had produced industrially relevant new results it appeared that, once again, firms in America were showing more interest and were more likely to exploit the output than firms in the UK.

Conclusions

The Animal Cell Biotechnology Club had a number of different objectives – establishing a critical research mass in this area, establishing a pool of trained researchers in this area, and the fulfilment of a number of specific research projects. The Club should have benefited more from economies of scope than scale (though shared samples and cell lines provided some scale economies) and from externalities. However, due to low interaction the scope economies were low and benefits of being in the network were ultimately related to funding not to network participation.

Although those involved in the Club considered it to have been valuable in some ways, there has clearly been a number of problems in the operation of this Club. These problems seem to arise from two inter-related aspects of the Club. Firstly, the role of the steering group and of the companies on the steering group. Secondly, the aim of developing a scientific area where previously there had not been scientific expertise. These two aspects meant that on the one hand scientists in the Club were being subjected to much more control, direction, and criticism of their research – especially, but not only, in the second phase – than they were accustomed to; at the same time there was potentially more scope for such criticism due to the fact that many scientists were having to learn a number of new skills in a new area.

These problems raise much wider issues about science and industrial policy. The problems experienced do raise important questions for the current science policy of the UK, in particular its emphasis on industrial relevance and the role of industrialists in managing and monitoring university research. It indicates the need for very clear specification of management and steering group responsibilities where these exist. It also indicates the need to ensure that medium and long run research aims are not inhibited or distorted by shorter run industrial aims. It also raises much wider questions about when or whether it is appropriate to have a science policy that is in effect an industrial policy. The science policy aims in this case were determined by the perceived needs of the biotechnology industry. The extent of beneficial industry effects remain to be seen. The precise value and quality of the scientific activity cannot be assessed in a case study of this kind, but there were strong criticisms by the scientific evaluators at the end of phase one. This suggests there is scope for debate and for further research as to when and in what ways funds should be allocated to projects where it is clear there are steep learning curves involved. This is a debate for both science policy and industrial policy.

6 Peroxidases in Agriculture and the Environment (Case Study 4)

This case study is of a network of scientists funded under the European Commission's Human Capital and Mobility programme. It is a relatively small network and at the time of the case study was in the first half of its work programme. The case study is of particular interest for a number of reasons: the science involved is highly interdisciplinary, the core group of researchers in the network existed prior to the Commission funding this network, and the network is one where human capital and mobility is emphasised.

Overview

The network was formally established in 1993. It has funding for two years, but the overall work is likely to be spread over a period of three years altogether. There are seven universities and research institutes involved in the project. Three of these come from the less favoured regions – Greece, Ireland, Spain. The other four come from the UK (two), Denmark and Italy. The four teams in the favoured regions already had contact with each other and were working together on at least a bilateral basis prior to this particular network. These institutions also already had contacts with the institutes from the three less favoured regions that were also included in the network.

The fact that the core group of institutes were already working together was of importance both in putting together the research proposal and in being awarded the funds – it was made clear in the proposal that the core of the network already existed and already had joint research experience. There had been a scientific meeting in London before the network was formed where people from Denmark, Italy and the UK had met and discussed their scientific work and where it had become obvious that it was possible for them to make significant contributions to each other's science. It was from that point, that this core group began to discuss the possibility of forming a formal European network. Although there were political considerations in establishing the network, in terms of making it a larger network including less favoured regions, nonetheless, the science was seen, in general, as coherent and not contrived, and the network did build on existing knowledge and contacts. There were some differences in views as to the contribution to the research the different laboratories would be able to make.

The fact that the project is part of the Human Capital and Mobility programme means that one of the aims of the network is not simply to produce new scientific results but also to provide training. This should be achieved by the different institutes employing scientists from within the EU but not from their own country. Whether the training element causes inefficiencies in the overall research programme was seen to depend on the quality of the applicant and on communication skills. It was not seen as being compromising to the work of the network in general.

The network's research is concerned with an important class of enzymes called peroxidases. The aim is to understand the structure, mechanisms, regulation and expression of peroxidases. The network is focused on horse radish peroxidase and is also undertaking a comparison with an insect peroxidase. Members of this family of enzymes are of major environmental and commercial importance with many varied potential applications. At the same time, much of the science involved is fundamental and basic in nature. The examples of potential commercial uses range from applications in washing powders to applications to detoxify the environment. The science is interdisciplinary involving molecular biologists, bio-chemists and physical bio-chemists. The science is interdisciplinary not only across the different institutions involved but also within those institutions. Thus, for example, the network is co-ordinated from the University of Sussex, and there is an interdisciplinary team there in different departments and institutes who had already worked together and are working together on this project. Since the work is interdisciplinary, the different research institutes in the project are not working separately on discrete projects but do have substantial interaction. Different institutes have different connections across the network, so that they will work more intensely with one or two of the institutes than with some of the others. Working with an interdisciplinary team with different types of expertise enables questions and areas to be focused on that would have not been possible otherwise.

Motivations

The major motivation for establishing the network under the human capital and mobility programme was to obtain funding in order to further scientific opportunities perceived in this field. The project built on research that was already running. It was clear from existing research interactions that there were important complementary groups in Europe. The need was for funding to enable this research and this collaboration to be built on and taken forward. Thus, the motivations for doing the science and for the collaboration already existed – there were strong benefits to be gained from the different interdisciplinary expertise of the members of the team. The European programme provided the opportunity to obtain additional funds both to further the research and, possibly to take it forward more quickly than would have been possible otherwise. Most members of the network were clear that they would have expected to continue this research through applying for alternative sources of funding if this particular project had

51

not been successful. Nevertheless, in one of the less favoured regions, the view was expressed that it may only have been possible to proceed with less than half of the research that is being undertaken as a result of the project. The core of the network would have functioned anyway in the absence of this grant – they would probably have applied to national research agencies or even to a different EU programme. Motivations in the less favoured regions included the opportunity to collaborate with leading groups in Europe. Participation in European networks is seen as providing some status and also as being seen positively within researchers' university environments.

Management and communication

The co-ordinator of the network has an administrative role to play in ensuring the successful functioning and operation of the network. He has an obligation both to ensure that the overall work programme of the network is fulfilled and that the training element within the Human Capital and Mobility programme is also fulfilled.

The network had applied for about 1.8m ECU and was awarded about 1/2m ECU. There were time lags in waiting for the EC programme and contract to come through. The scientists are working on a range of projects, and delays may mean that their research has developed in other areas or in related areas ahead of what was proposed within the programme.

There were two initial meetings of the network to discuss the functioning of the network and the distribution of the funds. The funds were primarily for employing people in the different laboratories. Under the terms of the EU programme, those employed had to come from other EU countries. There were some disagreements in coming to decisions on how to allocate the funds. This was partly to do with differences in the research strengths of the different institutes and differences in the emphasis that some of the scientists were willing to place on the training aspect of the network. Some problems also arose due to the different costs of employing scientists in different countries. Thus, a laboratory in a higher cost of living country may get more money than one in a lower cost of living country but actually be able to employ somebody on the project for a much shorter period of time. The money was allocated in such a way that all the laboratories were able to employ at least one post-doc and the stronger laboratories in some cases were able to employ post-docs for up to two years.

One problem that laboratories both in the favoured and the less favoured regions face is in actually appointing post-docs for the work. This may be due to salary problems, to a lack of appropriately qualified people coming through, and to a disinclination to be mobile. The UK is seen as an attractive location for post-docs because students can learn an international language at the same time as doing a science.

There is, inevitably, substantial scope for difference between the overall aims of a network in terms of training and mobility and the aims and interests of particular laboratories. Individual laboratories are concerned with the resources that they get out of the project. Especially where funding is substantially less than was initially applied for, participants may be less willing to give priority to the overall aims and needs of the network and are likely to focus more on their own specific needs and part of the project. In this network, this had not been a general problem, either in establishing the project or in agreeing the distribution of the money but it had caused some tensions and difficulties. Scientists felt relatively free in the way they could pursue their research within the network – two years was a relatively short period of time and so research was not going to change radically just for a two year project.

It is anticipated that the network as a whole will meet about three or four times over the life-time of the project. There is not a substantial travel budget and so more frequent meetings are not possible. However, there are other scientific occasions and international conferences where the network members are all likely to be in attendance and, where these opportunities arise, additional meetings of the network will be organised without requiring any additional funding. In addition, where one institute is working particularly closely with one or two other institutes, there are meetings between these different institutes, not necessarily using funding from the network. There is communication other than through the network meetings, because of exchange of personnel and exchange of samples. Some network research is driven by techniques which particular laboratories are specialised in and they can send samples to the other laboratories. There is substantial communication between the different institutes through telephone, fax and e-mail. Two years were seen as too short a period of time by some for a network, with three years as a minimum and five possibly far superior, given the learning and establishment costs.

National and international networks were not perceived as being particularly different. There was a view expressed that informal collaborations worked better than formal collaborations. The extent and nature of the collaboration was seen to depend essentially on the nature of the science and also on the particular personalities involved. Personalities were seen as a more important issue than more general questions of culture, nationality and language. Information exchange with people doing a similar type of research was a very important benefit of being in the network since reading the literature or attending conferences was not as effective. The interaction between laboratories was seen by some as closer than the interaction merely from knowing people in an international scientific community. Opportunities to collaborate with scientists in the US were seen as more limited as much as anything due to distance. The view was also expressed that EU funding is often additional and that there is less attention paid to duplicating of funds available on other projects whereas when the Americans collaborate with European partners they have to be very careful in terms of their American funding that the projects are distinct.

Some difficulties had arisen in management and communication of the network due to industry links. One of the participants had an on-going research link with a major bio-technology company. This link was referred to in the research proposal. Subsequently, the co-ordinator considered bringing in a different company from another country. There were sharply contrasting views as to the purpose and appropriateness of this being done. Where more than one company is involved in a network this can raise difficult questions for the scientists involved, in terms of the amount of information and expertise they are willing to contribute and exchange within the network.

Overall, there were not seen to be any major communication problems. This was partly to do with the relatively small size of the network. It was also related to the fact that the main members of the network already knew each other and had substantial experience of working together. The science they were involved in was perceived as being international and there were not seen to be any particular communication problems due to cross-country communication nor due to distance problems.

Outputs

Substantial scientific outputs were expected from the project. Scientific advances already had been made and many papers produced. The overall work of members of the network – not only on the particular project within the network – was seen as contributing to European science and giving it a high international profile in this area which was recognised by American and Japanese scientists. There was a view that there is now a critical mass as well as levels of excellence in Europe in the scientific area. Science in a single institute, relative to a network, is limited by the in-house expertise, by the expertise that can be bought in and especially by the amount of capital equipment that exists and the sophistication and expertise necessary to get optimum performance out of that equipment. In some cases, scientists had much closer relationships with people abroad than with people in their own institute – this is a question of scientific expertise and interdependency. The network raised the scientists' profile and gave them a higher profile at meetings and with other investigators and in that sense it had exposed their work to a wider audience. This had led to some further collaborations.

Participants were benefiting substantially from the different expertise of the members of the network and this was benefiting their science not only within the project but in terms of other projects and other possible future collaborations. The network provided a base for pulling in additional funds. Members of the network were involved on other collaborations and had substantial funds from other sources on related research. At the same time, the existence of the network did not necessarily mean that they would continue to collaborate in the future or that this would inhibit potentially valuable collaborations with other people in the future. In the short run, it inevitably meant that, even if other possible interdisciplinary collaborations would be possible for the same task, the network was used. This was not seen as particularly limiting. The initial contacts which existed were made stronger by the network and were likely to continue after the network has formally ended. Good collaborations will also have been established with the new laboratories. One scientist thought to some extent European funding did mean that scientists are sometimes not collaborating with the best people but rather with Europeans.

The interaction between the institutes in the favoured and less favoured regions is not by any means entirely one way. Differences in scientific development can have important implications. For example, Spain has a number of good enzymologists but fewer good molecular biologists. In contrast, in the UK, the focus on molecular biology has resulted in less focus on enzymology and so the Spanish can make strong contributions in this regard.

The network was not perceived as inhibiting competition between scientists. Furthermore, since science is interdisciplinary, the scientists in the network are not necessarily directly in competition as they would be if they were in exactly the same area of specialisation. Networks can act to limit duplication of research and to make research more complementary. It is at the same time important that there is still competition, but this can also be international with respect to America and perhaps Japan. A common understanding of the issues does arise but probably due to the dynamics of the science and not to do with the network per se.

The network did not have a major role in widening participants' contacts with other members of the international scientific community. This was because the scientists were already known to each other and in the central cases already working together. The network was simply reinforcing these interactions rather than creating them.

There were inevitably costs in terms of administration and communication in participating in the network. The EU was seen by some as particularly bureaucratic and increasing administrative loads. Where relatively small amounts of funding are received, then the time taken in establishing a network and fulfilling administrative requirements can seem excessive, especially where scientists may have relatively easy access to alternative sources of funds.

Conclusions

This network appears to be well constructed in terms of its scientific rationale and aims. This is a result both of the interdisciplinary nature of the science and of the fact that the scientists have already worked together in various forms. Although there had been some problems in initiating the work of the network, due to allocation of limited funds and the need to ensure a training element within the network, these do not seem to have been major problems and the network is operating effectively. It is fulfilling its training requirements and these are not in general seen as inhibiting or detracting from the scientific purpose of the network. There are not only major economies of scope from the network due to the interdisciplinary nature of the work but there are also important economies of scale since much of the equipment is highly specialised and expensive and by collaborating across institutions and countries different institutes can have access to one piece of equipment in one location. There were not seen to be any great advantages in bringing a whole interdisciplinary team together on to one site. Indeed, one participant commented that unless people are working together in one laboratory, there is very little difference between being 300 metres and 300 miles apart. This is then an example of a network that functions extremely well due to the nature of the science and where the EU funding plays an important role in an enabling rather than an initiating sense.

7   Critical Loads Advisory Group (Case Study 5)

This case study focuses on a British network in environmental science that has strong connections with the European environmental network described later. The network is of interest as it brings together scientists from diverse disciplines and organisations, and also brings scientists into unusually close liaison and communication with policy-makers at national government level.

Researchers were interviewed at four institutes participating in the network, and included members of the network's Steering Group. A senior manager from the Government Department funding the network was also interviewed.

Overview of the network

The Critical Loads Advisory Group (CLAG) is a British network of twenty-five organisations, including universities, other research bodies and public agencies, plus one private sector participant working in the field of environmental sciences. The CLAG network is concerned with the development of 'critical load' indicators for soils, water, vegetation and other environmental 'media' affected by acidification from atmospheric pollutants. The network also carries out work on development of methodologies for establishing such indicators.

A critical load is the level of exposure to acidification (or other pollution) below which there is no significant damage to the ecosystem or building in question. Where critical loads are exceeded there may be sustained harmful effects. Given the highly complex interactions involved in acid rain and degradation of ecosystems, this is a field subject to considerable uncertainty and scientific controversy. It is, like all fields of enquiry in environmental science, inherently multi-disciplinary because of the need to examine the interaction of many different geological, chemical and biological systems.

It is also a field in which, unusually, scientific findings can have an immediate impact on policy-making processes in a number of areas (for example, health, transport, industrial regulations, energy generation). The policy questions are in turn characterised by cross-departmental and multi-disciplinary issues. For these reasons environmental science is particularly reliant on collaborations between disciplines and institutions, and is notable for a high level of networking activity.

The CLAG is overseen and funded by the UK Government's Department of the Environment (DoE), and is co-ordinated by a public sector environmental research agency, the Institute for Terrestrial Ecology (ITE). As well as carrying out research at several laboratories and field sites, the ITE acts as the managing agency for the network, which is split into 6 sub-groups comprising representatives of the research bodies working on different subjects (soil, freshwaters, mapping and modelling, etc) and which form the nodes of the national network. The largest of the sub-groups, with some 26 members, is that concerned with mapping and modelling critical loads for the UK: this draws on the basic scientific research carried out by the other sub-groups and integrates the findings into models and computer-based maps (using Geographic Information Systems software).

The CLAG brings together researchers from public agencies such as ITE and the Meteorological Office as well as scientists from universities in the UK. Numerous parts of the ITE organisation are represented on the Steering Group and the sub-groups. The Department of the Environment and two other policy bodies (National Rivers Authority and the Forestry Authority) are represented also, along with one private sector consultancy. There is a mixture of types of organisations in all of the sub-groups and also on the CLAG's Steering Group, which sets the overall strategy for the network. Some of the researchers work in more than one sub-group, and a number are on the Steering Group as well as at least one of the sub-groups. Thus there is considerable overlap and a number of key individuals are able to gain a comprehensive view of the network's activities.

The aim of the CLAG is to carry out research in the field of critical loads that will result in the development of maps, models and indicators that can be used by policy-makers negotiating Europe-wide agreements on reductions in emissions of pollutants implicated in acidification damage to the natural and built environment. The research is thus directed towards the delivery of specific 'products' for the DoE, but there is also a need for less focused 'underpinning science', as one researcher said, in order to refine the critical loads concept and investigate impacts on environmental media.

The CLAG produces maps of the UK showing the sensitivity of different soils, plants and waters to acidification and the levels of acid deposition experienced in different areas. These maps complement mathematical models of acidification and critical load indicators of the sensitivity of different ecosystems in the UK to deposition of chemicals such as sulphur. The focus for future work will be critical load maps and models relating to nitrogen compounds in the environment. The UK work is not only for domestic use: it feeds into a wider European network on critical load modelling and mapping, which serves the needs of European conventions on transboundary pollution (see Case Study 6 below). The ITE acts as the co-ordinating member for the UK of this pan-European network, and other CLAG members are collaborating within the European programme.

Establishing the network

The need for a research programme on critical loads was identified by the DoE in the wake of controversies over acid rain pollution in the early and mid 1980s. The UK was implicated as one of the main offenders in western Europe on account of its high level of sulphur emissions. The DoE wished to support scientific research on acidification and critical loads to underpin domestic, European Community and pan-European negotiations on reductions in emissions. It was recognised by the Department that it could not possibly bring the necessary expertise to bear using in-house staff and resources alone. Moreover, it was felt that the development of a network of leading research centres would generate synergy and provide 'a measure of peer review' that would give the Department 'state of the art' research results with which to work.

The Critical Loads Advisory Group network was set up by the Department of the Environment in 1988. Initially there was a small advisory group, comprising key researchers identified by the DoE. The core membership of the network was built up from organisations with which the DoE was already familiar as a result of contracting research projects in allied fields, and which were acknowledged centres of excellence in the general field and in specialist areas. The network members thus had a shared link with the funding body, and were familiar with one another's activities through existing contracts and academic linkages.

As its goals crystallised and more areas for research were identified, the group built up more connections with the environmental science research community using members' personal contacts and recommendations. As the number of researchers grew and research issues were identified so the sub-group structure evolved, and eventually the structure was sufficiently complex and extensive that a Steering Group was established in 1991. This Group has thirteen members and is chaired by the DOE representative, with ITE providing the secretariat. Again, the contracting process of the DoE played a key role. The core members began to sub-contract pieces of work to other centres, which then were drawn into the CLAG in a formal sense through the sub-group structure.

Other bodies were brought in as existing members realised that they required new specialist expertise or extra assistance in certain areas of work. One original network member said that there had been a tendency to restrict the CLAG to sub-contractors and collaborators already known to the core organisations. While a few organisations outside these categories had been co-opted to sub-groups because of their knowledge and importance in policy terms, some researchers had been 'uneasy' about this – it was almost as if they were eavesdropping' on the research debate. A key anxiety was that including several bodies outside the specialist research community would risk 'diversion on to other issues' and 'political agendas': it was felt that it would be difficult to draw the line in asking new members to join the CLAG on the basis that they had a policy interest in the research issues and results. (This concern reflected a more general tension within the network, concerning the unusually close links with the world of policy-making and political negotiation: this issue is considered in more detail later.)

The organisation of the network in the UK thus developed in a gradual and 'organic' fashion, with no top-down imposition of ideas or structures from the DoE. Its form evolved in response to the developing research agenda framed by the Department. The network also evolved in parallel with critical loads work on the European level, in which DoE was involved as the UK Government representative at international negotiations on transboundary air pollution. However, respondents stressed that the research agenda has not been dictated by the funder in any detail. There is considerable autonomy for the various sub-groups to develop lines of enquiry and direct their programmes of research following strategic guidelines and priorities established by the Department and discussed by the Steering Group.

Motivation for involvement

The researchers interviewed were unanimous that a key motivating factor was the availability of funding from the DoE. Funding benefits were felt to be significant in terms of research salary costs and also purchase of equipment that otherwise could not have been obtained. The researchers all felt that the work they were doing could not have been funded on the same scale without the DoE's involvement, or that their personal involvement in the critical loads field would have dwindled or ended altogether. The researchers all noted that the availability of public funds for academic work on pollution and 'near-policy' issues was relatively restricted and that there was increasingly intense competition for grants for academic research. Given this background, there was a strong motivation to seek funding from such a major Government-funded initiative. The funding also made it possible to take on more research assistants and to open up new opportunities in research for graduate students: one CLAG member estimated that around twenty PhDs had emerged from the programme as a spin-off effect, as researchers steered students towards CLAG-related topics. A similar number of Master's degrees was thought to have been generated in the same way.

In addition to the contribution the funding made to salary costs, equipment and opportunities to take on more research staff to assist projects, there was the potential for 'spin-off' contracts from other funding bodies. The expertise developed in the course of the CLAG work was attractive to other public agencies concerned with pollution issues, and the connection with the CLAG network could be a strong 'selling point' in securing further research contracts.

Spin-offs were also reported in a less commercial sense. The insights gained carrying out the CLAG contract work could stimulate new lines of enquiry for more purely academic research, and also lead researchers into new areas that they might otherwise not have considered. It was reported by all the interviewees that they had developed many new contacts with other teams within and outside the network as a result of working on the CLAG programme. While some of these contacts might have been made in any case, it was felt that, in the absence of the

CLAG, opportunities for such networking would not have arisen so often and would have been more a matter of chance. In general, there was a strong view that participation had contributed to a less insular attitude among researchers, both in relation to links with other institutes and to exploration of new areas of research.

As well as funding opportunities and a range of contacts and spin-off benefits, there were substantial intellectual benefits for network participants. All of the interviewees said that the CLAG was attractive to them because it allowed them to work in fields in which they had already developed interests and expertise and brought them into contact with leading researchers elsewhere in the UK, and indirectly elsewhere in Europe. This broadening of horizons also could have beneficial effects on research and teaching more generally within the organisation.

It was argued that although the DoE sometimes had very specific priorities and urgent needs for answers, there was substantial scope for the research to proceed in a traditionally 'pure' academic fashion: researchers felt that they had considerable autonomy despite the contractual nature of the link with DoE and the applied and strategic nature of the programme. One researcher said that while the mapping and modelling side of the CLAG operated under some pressure, DoE managers recognised the extreme complexity of the other sub-groups' areas and allowed them considerable leeway in setting their own priorities. The initiative was also considered satisfying in that it brought coherence at a national level to the area of critical loads, and linked the efforts of disparate groups to create a climate in which the 'whole could be greater than the sum of the parts', as one researcher put it.

A further motivating factor for some was the policy dimension of the network's programme of research. As noted earlier, the CLAG's work is unusual in that its results have a bearing on political negotiations, in this case on emission of pollutants in Europe. For some of the scientists involved, this dimension of the network provided an additional motivation to participate. For a few, it gave the sense that one could play a part, however indirect, in improving environmental quality and affecting policy-making for the better. For others, there was a more general interest in seeing how policy-makers made use of scientific findings and advice. As one researcher put it, 'you have to be realistic... it's hard to tell how much notice policy-makers take of what we do', but nonetheless the CLAG network's contact with the DoE provided 'an interesting opportunity to see just how much notice is paid [to science] in this kind of context'.

The political dimension of the research network was, however, a factor that apparently deterred some scientists from taking part, or which some participants wished to avoid as much as possible. One interviewee said that some researchers outside the network took the view that participants had become too closely linked to the agenda of policy-makers and had thus 'sold their birthright' as scientists. Others acknowledged that at times there were pressures from policy-makers for rapid results and responses, and for clear-cut answers wherever possible. Given the complexities and uncertainties in this area of environmental science, this could not be done without a degree of over-simplification and coarseness in the presentation of findings. This was inevitable given the complexity of the environmental science involved in critical loads and the long-term nature of the research, and the comparatively short-term timetable of policy-making and negotiation. However, while some researchers were willing to accept that aspects of the CLAG research might be seen as 'dirty science' – rough and ready applied work – that they could refine and pursue in a 'purer' fashion outside the CLAG framework if necessary, clearly some outsiders felt that the work was compromised and simplistic because of the close connection to policy.

Thus the essential difference seemed to be between scientists willing to trade off the 'quick and dirty' aspect of some of the responsive work done for DoE against the powerful attractions of funding, contacts, spin-off benefits and scope for influencing policy, and those wishing to retain control over their own 'pure science' agenda. Some of the latter had evidently found a place within the CLAG network but were at pains to keep their distance from the policy side; others were outside the network entirely. For those who wished to develop their skills in communicating their work effectively to policy-makers, there were evidently opportunities to 'carve themselves a niche' in policy-related science, as one CLAG member put it.

A further motivating factor for CLAG members was the recognition that professional progress in their field was increasingly linked to the ability to work in collaborative networks. Environmental science is inherently multi-disciplinary and demands collaboration across specialist boundaries if conceptual and empirical progress is to be made. As one researcher said, this leads scientists to seek out collaborators, and those most suitable are not necessarily to be found in one's own institute. The field demands linkages between centres because researchers need access to one another's data and wider perspectives on research priorities, and because so many different specialisms must be brought to bear on the problem of critical loads for any given interaction of a pollutant with an ecological medium. The network benefitted in particular, therefore, from important economies of scope. However, scale economies also existed – both in terms of specialised equipment and the overall scale of the project.

Respondents all agreed that the 'group science' mode of work was increasingly important: as one said, 'It's a fairly common problem, now that group science has taken over, that individuals ploughing their own furrow find it more difficult' [to secure funding]. They expected that ability to work in collaborative networks would be a critical factor in career progression in future. One interviewee went so far as to suggest that it was becoming something of a 'stigma' not to be involved in collaborative initiatives at national or international level. Another senior researcher from CLAG said, 'My life is dominated by putting together national and international consortia'. In relation to this, it was felt that participation in the CLAG network would be beneficial when putting in applications for further research to public agencies in the UK or to the European Commission: it would be a good indicator of capacity and aptitude for collaborative work.

As far as the sponsoring organisation was concerned, the motivation for establishing the network was clear. The DoE could not handle the amount of work needed in-house, and the network mode provided 'a degree of peer review' and brought together the leading organisations in the field. Moreover, many were already known to, and trusted by, the Department. The work was thus evidently 'state of the art' research and the range of academic and other research organisations involved conferred authority and independence on the programme.

Management and communications
In the CLAG structure, the Steering Group defines the areas of work to be covered and the sub-groups elaborate specific programmes of work within their sub-disciplinary areas in order to meet the overall goals. Within the sub-groups, members' teams work on distinct but complementary areas of research and develop research plans and proposals for DoE 'as a group of collaborating researchers', as one sub-group member said. On the basis of these discussions bids are submitted to the DoE for new allocations of funding or requests are made for use of existing allocations. Many groups are also involved in related work not funded by the DoE for the network: as a result of their involvement these groups often re-orient existing related research towards the interests and priorities of the network. This provides a potentially useful spin-off for the ITE as co-ordinator and for the DoE as funder of the CLAG network.

The operation of the network was considered to be fairly relaxed in terms of DoE control of the agenda, but timetables can be tight and there is a keen awareness among researchers of the need to produce 'deliverables' on time. As noted earlier, the mapping and modelling role is subject to rather stricter deadlines and project management than the other areas of substantive research, on account of the direct link between the critical load maps and the needs of policy-makers and negotiators. Moreover, the mapping work produces outputs that are also required by the wider pan-European network described in case study 6 below.

The Steering Group members have between them considerable experience of working in 'group science' and of managing national and international collaborations. Some have heavy network commitments in parallel with the CLAG: for example, in developing multi-disciplinary and multi-institute programmes of research on global environmental change. The ITE is particularly accustomed to managing networks, on account of its involvement in many areas of environmental science and the need to co-ordinate work across its numerous UK research centres and field stations.

Thus there is a high degree of familiarity with the network style of operation and the core members of the CLAG have developed the specific management and communication skills needed for collaborative ventures. It was noted by one network member that the CLAG was fortunate in this respect: in other disciplines, funders and co-ordinators of networks might need to set aside resources for training scientists in the managerial and communication skills required for good collaborative networks, and for communication with policy-makers where there was a need for this.

It was felt that in some institutes there were individual researchers who resisted the network approach: some of these could be 'steered' into productive work, while others needed to be 'dragged', as one CLAG member put it. However, the intellectual and financial incentives and pressures to engage in collaborative networks were such that there was relatively little resistance to participation as far as environmental scientists were concerned.

As noted earlier, the original network structure has been made more formal as the programme of work and the number of new members have grown. For the DoE and the ITE the Steering Group and sub-group structure balances the need for a high level of academic autonomy with the need for strategic direction and quality assessment and for liaison with the Department's policy-makers. The overlapping memberships of the Steering Group and the specialist sub-groups ensure a sufficient flow of communication between the sub-groups and the management level. As one member noted, sub-groups within such networks 'tend to take on a life of their own', and could be tempted to 'spin out' research funds for as long as possible, whereas the funding body required relatively quick responses and reporting of results. With the structure that had been developed, these tensions seem to have been minimised.

One member noted the potential problem that networks such as the CLAG could be seen simply as a means of 'horizontal' linkages between senior researchers, collaborating in a club of 'heads of labs'. It was important to ensure that the more junior researchers doing much of the practical work were also integrated into the meetings. This 'vertical integration' was done through meetings of the sub-groups, and provided a means of training post-doctoral staff and others in network operations and allowing them to take on more responsibility.

Meetings of the Steering Group and the sub-groups take place 3-4 times a year, and the location varies with different CLAG member organisations acting as host. In addition, informal bilateral and multilateral meetings may take place. Meetings cover issues of network management, deadlines and timetables for delivery of data and reports, research problems, and so on. In addition, there is an annual meeting of all of the network members to review progress and 'brainstorm' about future priorities for research. The formal meeting structure brings in a measure of 'continuous assessment' of quality of management in the network, in the view of one respondent.

A special full-network meeting was also held in 1994 to consider management of the CLAG in the light of the imminent retirement of the DoE's representative, who was Chair of the Steering Group. The network will also need to reconsider its structure in view of the forthcoming shift in research focus from sulphur compounds to nitrogen impacts in the environment: this reflects the changing needs of European and UK policy-makers, and may require recruitment of new members to CLAG and a reorganisation of sub-group specialisms.

Outside the structure of regular and ad hoc meetings, most of the communication is by telephone. Network members are also likely to come across one another in other fora, such as academic conferences on related environmental research topics such as climate change. For data exchange fax and e-mail are used; e-mail via the JANET universities computer network is used for transfer of critical load map data held on computer in Geographic Information Systems format. Occasional problems in transferring data and the lack of a common CLAG database were mentioned, but these were not regarded as serious difficulties. (Moreover, by late Spring 1994 the ITE had been given the task of setting up and operating a central database for the network.)

Tele-conferencing has been discussed as a possibility for the future but is not a priority: one researcher felt that applications might be 'few and far between' – the set-piece meetings are major decision-making affairs and members would be unlikely to wish to substitute tele-conferencing for face-to-face contact in this context. In addition, there is the issue of uneven spread in infrastructure to support tele-working among all of the network members.

Overall, the key difficulty in communications has not been technical issues, but rather the problem of finding common dates for meetings among hard-pressed researchers. The main difficulty for ITE as coordinator of the CLAG network has been bringing people together for ad hoc meetings at short notice on specific issues: the problem is that members are increasingly involved in meetings on other issues (typically concerning organisational change in their institutes) and find it difficult to attend sessions for CLAG at short notice.

In general, CLAG members felt that the management and communications of the network (and of others in which they were involved) were not essentially different from, or more problematic than, the operation of a large multi-disciplinary initiative in a single institute. The main difference was that in the latter case there would be more opportunity for day-to-day discussion of issues and members would be more up-to-date with news and aware of forthcoming activities of colleagues in different areas of the research programme.

Problems of differences in organisational culture did not loom large: it was felt that these could arise just as easily within institutes as between them. It was also felt that no particular problems had arisen in terms of skills and training for effective participation in the CLAG. The main need of network members was for 'acclimatisation' or 'conversion time', as one researcher put it, to adjust to the workings of the system, rather than for specific training.

Outcomes of the initiative
All of the interviewees regarded the CLAG collaboration as a successful exercise, both in terms of network management and quality of scientific work. The work

was demanding, and there was the additional pressure of responding at times to policy needs, but all felt that participation was very worthwhile. One described his involvement in the CLAG as 'a turning point in my career', opening up new research horizons and leading to opportunities to communicate and work with scientists and policy-makers in the UK and the rest of Europe.

It was felt that the research teams had gained considerably from participation: the CLAG had given them secure funding for research that otherwise might not have taken place or that would have been much less systematic, and there had been numerous spin-off benefits (contracts, access to equipment, new academic contacts). The funding body had obtained a steady flow of high quality work that met the needs of policy-makers and advanced understanding of the critical loads concept and its application to ecosystems in the UK. The CLAG has produced some twenty research papers per year on average.

Formal evaluation of the network has not yet taken place in order to assess the value for money or performance against targets. However, the scientific outcomes of the work were seen as satisfactory and reflecting the 'state of the art' by participants. It was felt that there was a high degree of additionality in the programme: without the funding, all agreed, similar work would either dwindle into insignificance or take place at a much lower level of funding and consistency owing to constraints on other sources of research funds.

Respondents felt that most of the objectives of the work programmes are being met without major problems. Network operation seems to have functioned smoothly, and the structure of the CLAG has evolved in a way that allows the research groups considerable autonomy while ensuring that there is clear strategic direction and liaison between sub-groups. It was noted that the CLAG exemplified good practice in key respects:

- it is not a 'passive' network simply for information exchange, but is action-oriented, with clear research goals;
- the mutual benefits are clear for the funder and the research teams, and participants have a common interest in meeting each other's needs;
- meetings are geared to practical action and are regular, promoting cohesion in the network and efficient flows of information.

The general approach taken to network design was regarded as fruitful and an instance of good practice for collaborative scientific networks: first the network was allowed to evolve as seemed most suitable for the core members and the needs of the programme; once problems of size and communications arose as the network expanded, a more deliberate process of network design took place and the group structures were formalised. This approach appears to have avoided early rigidities and made later adaptation easier, and has also built in a suitable degree of flexibility and overlap between groups.

A few reservations concerning the network were noted, however. First, the network structure has meant that reporting findings and collating data have become more complex operations than normal because of the large number of network members and the different interests of the agencies involved. The quantity of data to be handled for the GIS mapping work is probably much larger than would be generated by any single-institute project. This led to some frustration: one researcher said that the timetables for the CLAG work meant that there was insufficient time to analyse and interpret the vast amount of data collected. The datasets represented a huge scientific resource and he felt that full justice could not be done to them within the framework of the initiative.

Second, some members found that the costs of attending meetings and carrying out the necessary administration involved in networking were higher than they had budgeted for. It was emphasised that care was needed on the part of funders and researchers alike in allocating 'realistic' sums for travel, subsistence and the administration of network participation.

Third, some respondents agreed that there was a potential problem of restriction of the overall field of research because of the concentration of leading centres in the CLAG network. One said that it was a possible danger that the network could become 'too incestuous' and stressed the need to bring in 'new blood' at intervals. The shift in focus from sulphur impacts to nitrogen would provide an opportunity for rethinking structures and membership. A related point was made by another interviewee, who noted the danger that network members could be drawn into the CLAG work as if on a 'treadmill', and could find that the network crowded out opportunities for other research they wished to do.

Another researcher said that there was reduced competition 'without a doubt' because of the breadth and scope of the CLAG network: those outside might well feel cut off, and it was clear that network funders and co-ordinators needed to be sure that they had included the most suitable institutes in the network. Efforts were made to disseminate the results of the CLAG research widely in the research community, through seminars, conferences and publications: findings were not kept within the network. Respondents also agreed that it was good practice to ensure that network members drew on ideas generated by researchers outside the CLAG.

In general, the CLAG network was regarded by all respondents as a successful exercise in collaboration that offered pointers to good practice to other disciplines. Benefits arose in particular due to scope economies, but also due to scale and externalities. Environmental science, by its nature multi-disciplinary, provides fertile ground for the development of networking skills and cross-disciplinary and cross-institute collaborations. The CLAG model appears to have struck a successful balance between the needs of policy-makers and scientists, and to have promoted a high level of fruitful collaboration that has generated numerous benefits for all members.

8 Mapping Critical Loads for the UN Economic Commission for Europe (Case Study 6)

This case study examines a pan-European network in environmental science. The network serves the United Nation's Economic Commission for Europe (UN-ECE), contributing data and maps of pollution impacts to support the UN-ECE's Convention on Long-Range Trans-Boundary Air Pollution. It is overseen by a Co-ordination Centre for Effects (CCE), which manages research on 'critical loads' mapping for the UN-ECE Convention's Executive Body.

The CCE network brings together scientists from diverse disciplines and organisations within the European Union, other west European countries and also countries in Central and Eastern Europe (including Russia). It also brings scientists into close contact with policy-makers at national government level.

Researchers were interviewed from four Continental institutes participating in the network to varying degrees, and included members of the network's co-ordinating centre. Views on the CCE network were also sought from researchers interviewed for the case study on the UK's Critical Loads Advisory Group (CLAG) described in chapter seven above: the CLAG is linked to the CCE system. Countries covered among the network members were: Germany, Hungary, the Netherlands, Poland and the United Kingdom.

Networks in environmental science
As noted in the CLAG case study above, environmental science is an intrinsically multi-disciplinary area: progress is dependent to a considerable extent on collaborative ventures within and between institutes and countries. As awareness grows of global environmental problems and of the 'transboundary' effects of much pollution, so the need for, and scope of, environmental research networks will increase. Problems in environmental science call for collaboration in a number of ways:

- ecological systems and the stresses placed on them by pollution are of extreme complexity and may be subject to dynamic change. They can only be understood if insights and data are brought to bear from numerous disciplines (including branches of biology, soil science, atmospheric chemistry, toxicology, etc);

- there is a critical need for scientists to collaborate in order to understand the science of interactions between different environmental 'media' (soil, freshwater, air, stonework etc);
- the vast amount of data gathered in the course of environmental studies demands close collaboration between scientists and information systems experts in order to extract as much value as possible from computer analysis and representation of data;
- in order to achieve international agreement on the extent of environmental problems and means of tackling them, there has to be close liaison between researchers to establish common definitions and standards for measurement of pollution;
- there is an unusually close link between fundamental science and public policy in fields such as environmental toxicology, monitoring of pollution damage, and global environmental change (for example, potential climate change induced by human activities). Environmental indicators are the essential raw material of political negotiations within and between countries on reductions in polluting emissions and associated changes in policy and technology. Collaborative networks thus bring together scientists not only into contact with their peers from different disciplines but also with policy-makers in a variety of areas.

The UN-ECE's programme of work to support the Convention on Long-Range Transboundary Air Pollution is one of many pan-European collaborative research initiatives in the field of environmental science. As well as the UN-ECE, several other international bodies co-ordinate environmental collaborations – the European Commission, UNESCO, and the European Science Foundation (ESF) are examples.

In addition, international networks based on bi- or multi-lateral collaborations are common. The UK's Institute of Terrestrial Ecology takes part in the UN-ECE research programme, and lists in its literature thirteen other European collaborative programmes with which it is involved – and this list is not exhaustive. Some of the major European networks are shown in the box below.

Some European networks in environmental science

CONNECT – network of conservation research centres

CORINE   – EU network for coordinated information management

ENCORE – European network of catchments for research on a range of ecosystems

ESF   Programme on Environmental Toxicology

UNESCO   Man & the Biosphere Programme – European network within a global initiative

STEP   – network for assessment of measurement techniques relating to forest decline

---

Overview of the UN-ECE system for collaborative research

The Economic Commission for Europe (UN-ECE) is a sub-grouping of the United Nations, covering member states across the continent. The USA and Canada are also members. Under the auspices of the UN-ECE, European governments negotiated the Convention on Long-range Trans-boundary Air Pollution of 1979.

This treaty concerns the development and implementation of international and national strategies for controlling and reducing air pollution carried across national borders. The treaty system provides for the negotiation of further protocols committing signatory governments to new measures and targets for reduction of polluting emissions.

The Convention is managed by an UN-ECE Executive Body, which oversees an extensive structure of subsidiary organisations. These are concerned with:

- development of international protocols for the reduction of polluting emissions;
- furthering cooperative research and information exchange on issues of air pollution control policy and technology, the effects of pollution on different ecosystems, and properties of different pollutants.

Critical loads research is crucial to the work of the Convention: it provides the scientific underpinning for the development of the protocols committing signatory countries to further reductions in transboundary air pollution.

The UN-ECE Executive body also promotes links for research and information exchange with other bodies in Europe, the USA and Canada concerned with environmental policy (notably the Commission of the EU). Links are also developed with countries outside the UN-ECE area that are dealing with similar problems of airborne pollution.

The UN-ECE Executive's subsidiaries include a set of Task Forces: these are networks of collaborating centres in member countries, co-ordinated by a lead country, which report to UN-ECE Working Groups on Effects, Strategies and Technologies. The network considered in this case study concerns the mapping of critical loads for transboundary airborne pollutants in Europe.

As explained in the CLAG case study above, critical loads are indicators of the concentration of pollutants in a given environmental medium below which there is no lasting damage to its quality and functions. Where critical loads are exceeded – for example, in relation to sulphur levels in freshwater – there may be serious or even irreparable damage to ecosystems. Key tasks for environmental science are to monitor the levels of pollutants across Europe in different media, to relate these to critical load estimates for each interaction of pollutant and environmental medium, and to produce maps to show critical loads and areas where these are exceeded.

In the UN-ECE Convention system, the Task Force on mapping of critical loads reports to the Working Groups on effects and strategies. It builds up its information from a pan-European network of research centres which are charged with mapping and monitoring critical loads in their respective countries.

This network consists of a set of 'National Focal Centres' – leading organisations in environmental research – and their domestic collaborators. The focal centres are a mixture of public research agencies; their partners are a mixture of public institutes, universities and ministry bodies.

The focal centres co-ordinate national efforts in critical load mapping and are the national point of contact with the rest of the network. Their activities are overseen by a Co-ordination Centre for Effects (CCE), based in Bilthoven in the Netherlands at RIVM, the Dutch national institute for environmental sciences and technology.

The CCE co-ordinates the network for critical loads mapping and disseminates its output to the Task Forces and Working Groups. The CCE's mapping network of national focal centres and their partners includes countries from all parts of Europe:

- European Union members: eg Germany, Netherlands, United Kingdom;
- Other Western European states: eg Austria, Finland, Norway, Sweden, Switzerland;
- Central/Eastern Europe: eg Poland, Russian Federation.

In addition, other European representatives of countries participate in the Task Force on Mapping and take part in workshop discussions with researchers from the national focal centres: for example, Bulgaria, Czech Republic, France, Hungary, Latvia, Lithuania, Slovenia, Spain. Links are also made with researchers in North America, China and other regions concerned with problems of transboundary air pollution.

Establishing the network
The concept of critical loads was developed by scientists and policy-makers in Europe and North America in the early and mid-1980s. It became part of the conceptual framework for the policy-makers and researchers working under the UN-ECE's Convention on Long-range Transboundary Air Pollution. After two workshops on the establishment of critical load measures early in 1988, it was decided by the Executive Body for the Convention that a Task Force on mapping critical loads should be set up. This was done with the Federal Republic of Germany as the lead country.

The Task Force was required to carry out detailed planning and co-ordination of research work for mapping the direct environmental effects of major airborne pollutants – such as sulphur dioxide and nitrogen dioxide, implicated in 'acid rain' damage to forests, freshwaters and buildings. The priority was to identify areas polluted by key chemical compounds, map them and develop methods for assessing potential damage.

The Task Force met in Germany in November 1989 and it was decided to establish a network of national focal centres that would produce critical load maps for their countries, drawing on the expertise and data of other domestic institutes as appropriate. The work of the national centres would be overseen and co-ordinated by a Co-ordination Centre for Effects (CCE), established in 1990.

The role of CCE was taken by the Dutch National Institute of Public Health and Environmental Protection (RIVM) in Bilthoven. RIVM volunteered to take on the work. RIVM is a major centre for environmental research, and the scientists concerned had experience of other European and American collaborative networks for work on acidification and on complex scientific modelling.

The national focal centres were not selected by the CCE or the Task Force. They emerged in a process of domestic discussion, with nominations made by the national governments. The focal centres needed to have a broad overview of the environmental science field, specific knowledge of the critical loads concept and techniques for applying it, and links with policy-makers and with the academic community.

This set of criteria narrowed the field considerably, and favoured major publicly-funded institutes or agencies linked to ministries. In the UK, for example, the national research centre ITE (see chapter seven above) was an obvious contender because of its network structure, academic and policy contacts, and role as the focal point of the UK's domestic collaborative initiative on acidification and critical loads.

The CCE thus had no control over the selection of the focal centres that it would co-ordinate: the national sub-networks coalesced in a self-selecting way. All national contributions are voluntary. The countries involved from the outset and at a high level of activity have been those most seriously affected by acidification problems through exceedance of critical loads – for example, Germany, Norway, the UK.

As the programme of the CCE network has developed, more members have been added, most notably from the ranks of newly democratic Central and East European states but also from Southern Europe. The former have long suffered from severe air pollution and also 'exported' it from their own industrial areas to neighbouring countries.

Relative newcomers on the Western side include Spain, previously one of the 'less committed' countries on environmental policy, as one scientist put it, but whose industrial development has led to acidification problems similar to those that have been familiar for decades elsewhere in Europe. The more experienced membership is able to provide advice and assistance to less experienced and less well-equipped institutes from these regions. The level of contribution to the overall research programme naturally varies considerably according to the expertise and resources of the countries concerned. Research institutes in the new democracies of Central and Eastern Europe contribute to the scientific and policy work, but they are also in need of extensive 'hand-holding' by the well-established institutes in the better-off countries that formed the original membership of the network. As one Hungarian scientist put it, inevitably 'some members are more equal than others' because of the different levels of expertise and experience in managing collaborative work.

Further extension of the network among Central and East European countries is probable. Potential full members such as Latvia and Lithuania are now involved in workshops organised by the CCE. Entry into the network will tend to depend on the capacity of the countries to make coherent inputs to the CCE and maintain consistently good quality (with suitable assistance from the better-off member states and more experienced research centres). This is likely to mean that those countries that are making a smoother transition to democratic order and a market economy will become full members of the CCE network sooner than those experiencing more political and economic instability.

Motivation for involvement

The motivating factors behind participation in the CCE network were common to all interviewees:

- availability of funding for critical loads research;
- the opportunity to make new contacts nationally and above all internationally;
- the possibility of influencing policy-makers;
- opportunities to develop new skills and enter new areas of research.

The weight attached to these factors varied from country to country. It was clear, however, that all were felt to be strong motivating forces for involvement by our respondents. Each is discussed below.

The availability of research funding from national governments to support work feeding into the Task Force on Mapping was an incentive to institutes to get involved. The UN-ECE itself does not provide research funds for participating centres. For some researchers the public funding from their national government made a major difference to their capacity to work in the field of critical loads. This was the case in the UK, for instance (see chapter seven above), where researchers felt that without the public money for the critical loads mapping work a coherent programme could not be sustained from other sources. A Hungarian respondent shared this view: given Hungary's difficulties in making the transition to the market economy, critical loads research was not a priority for Government and business, and so every means of alternative funding had to be sought.

In addition, all respondents took the view that future security in research lay in developing the expertise to work effectively in collaborative networks. It was felt that expertise in 'group science' and in gaining access to European programme funding were key skills for the rising generation of scientists. As one scientist put it, 'a team leader nowadays needs to be a better manager than a scientist'. This recognition of the need to acquire and update networking and managerial skills was a further motivating factor for taking part in the CCE network.

The potential for spin-off contracts as a result of gaining expertise and contacts through the network was also an important issue. Although the CCE network did not offer any more than marginal funding beyond what was provided by national governments, it could serve as a training ground for organisations wishing to win grants from international programmes that had large-scale resources for research. For example, institutes with little experience, as one respondent put it, of 'filling in those big forms' for major European Commission programmes, could learn a great deal from being in the CCE network. In this regard, CCE managers noted that they were 'inundated' with requests from researchers across the former USSR (now the CIS) asking for assistance in gaining access to European Commission programmes of research.

As well as the prospect of gaining access to other international programmes, the possibility of making new and valuable contacts was a key attraction for members of the network. The operation of the network gives organisations many opportunities to communicate with their peers in other countries and develop spin-off bilateral or multilateral initiatives. As one researcher said, it is 'a very good point of departure to get in touch with other institutes and gain a good overview, and to make profitable contacts too'. A related factor mentioned by scientists was the prestige and insight gained from working with leading researchers and institutes with international reputations.

This was welcome for respondents from all countries, but it was a particularly powerful motivating factor for researchers from the new democracies of Central and Eastern Europe. They had been deprived of free contact with their peers in the West for so long that the opportunity to join international networks was seized with special enthusiasm. One Polish scientist noted that the opening to become involved in the CCE network came at the same time as Poland and the rest of the old Soviet bloc was turning to democracy and the market economy, and was seen as a sign of liberation and a 'new era' in Polish science: 'We had the opportunity to become European... I had the chance to meet people I only knew from the literature'.

Contacts were not confined to European counterparts. The CCE network also brings researchers from other parts of the world into contact with the critical loads programme. The opportunity to meet experts from outside Europe was an additional attraction for the participants in terms of information exchange and the possibility of involvement in new bi- or multilateral initiatives.

As noted earlier, the critical loads field involves not only fundamental and applied scientific research but also the translation of data into a form suitable for policy-makers to use in shaping further work under the UN-ECE Convention. This degree of proximity to negotiations and policy-making is highly unusual in science. As mentioned in the case study of the UK's Critical Loads Advisory Group, this closeness to political issues is problematic for some scientists. It was stressed that participants can never forget that they are producing outputs that are of direct relevance to policy, as the timetable for delivery of data is largely shaped by the deadlines of the UN-ECE Convention policy-makers.

For some researchers, the policy dimension is evidently an attraction. One scientist working within a ministry felt that it was one of the main personal reasons for involvement: researchers could continue with basic scientific work and also see it being put to use in the policy arena in order to produce policies that would improve the environment and the quality of life. Finally, the CCE network was seen by respondents as a means of entering new areas of research and learning new research techniques (for example, in computer-based modelling and mapping) and methodologies. Critical loads methodology is a dynamic field and there is considerable scope for adaptation of existing methods and development of new approaches. The CCE was convinced of the benefits on all sides of bringing in less experienced partners: although '20 per cent of the people do 90 per cent of the really important work', the arrival of new institutes could reinvigorate the whole network. It was argued that 'you really fall asleep' if networks are restricted to a core team of expert institutes – 'we would never know what we were missing' if this were the case with the critical loads network.

The possibility of learning new skills and subjects was especially attractive for the East European networkers, for obvious reasons. One of the UN-ECE's aims is to use the network to promote the transfer of skills and knowledge to institutes in the new democracies. One Hungarian scientist stressed that this opportunity for learning and importing new techniques was far more important to him than the transfer of Western experts and equipment. A Polish researcher emphasised that participation in international 'pioneering' work on research methodologies was a major incentive for him – it had been 'impossible in the past' to have such involvement.

Management and communication

The CCE co-ordinates the network through regular communication with the national focal centres, which in turn manage their own domestic networks of collaborating institutes. The CCE has a number of key tasks:

- to liaise with all of the focal centres in order to ensure that national data collection and analysis standards are compatible with one another;
- to synthesise national data on critical loads;
- to provide guidance and training to researchers in the national institutes; and
- to oversee the production of critical load maps for each country and each pollutant and environmental medium, as required by the Task Force on Mapping.

It also serves as a clearing house on critical load data for the UN-ECE's Convention Executive and all parties to the Convention. The national focal centres must organise training in critical load mapping for their collaborators, collect and archive data according to network standards and prepare national maps for the CCE. Some countries are as yet unable to submit national data, and in these cases the CCE has the task of developing methods and data for them that will allow production of national critical load maps.

The co-ordination Centre at RIVM in the Netherlands is funded by the Dutch Government and RIVM itself. Government funding for the CCE is approximately Dfl 500,000 per annum. The Centre has three members of staff from RIVM to manage the network on critical loads mapping. In addition, the CCE pays for the travel costs of network members from the new democracies of Central and Eastern Europe for travel to workshops and other network meetings.

The network has been tightly managed from the outset, in order to ensure that outputs are delivered on time to the Task Force on Mapping to inform the policy making process of the Convention Executive. Timetables for critical loads work are governed by the negotiating timetables agreed by UN-ECE member countries: these are set out in an annual work plan approved by the Convention's Executive Body. The breadth of the network and its expansion, the variety of contributing institutes, and the need to achieve consistency of methods, data formats and definitions, all reinforce the need for rigorous network management. The extension of the network to bring in new countries has meant that over 70 people were involved directly in work with the CCE by early 1994. Network members interviewed felt that the management was very strict, and that timetables were demanding; on the other hand, it was evident that they thought the benefits outweighed these considerations. At the same time, CCE, management was co-ordinating across the national focal centres, while the latter co-ordinated within their countries. Thus, management of the research operated at various levels – not all was tightly managed from the centre. One complaint from a scientist at an East European institute was that the UN-ECE, like the European Commission in his view, was too slow in responding to proposals and other funding requests. However, this was not a complaint directed specifically at the CCE network.

The CCE managers said that the whole network benefited from a strict approach to timetables and project management – 'momentum' was created and participants were stimulated to react to new ideas and to do things that perhaps otherwise they would not accomplish. The CCE, for example, saw itself as setting the pace for discussion about methodological approaches that were new and controversial. Moreover, CCE managers said, the very proactive and focused style of network management meant that 'everyone feels involved' and the network has a group identity – it is not simply a loose collection of institutes with common interests.

Early in the development of the network it was decided to create a standard manual of procedures for the network members to follow. The manual is periodically revised, and sets out standard definitions of terms and outline procedures for building up critical load maps. The manual is prepared collaboratively by the CCE, the Task Force on Mapping and the UN-ECE's Secretariat.

The management of the network demands regular communication between the CCE and the national focal centres. English is the *lingua franca* of the network, and this was not felt to be a problem by any of the participating countries. Rather, it was seen as one of the essential skills for modern 'group science' and international collaboration.

Telephone, post and fax are the most frequently used methods, with electronic mail increasingly significant for map data exchange and delivery of training material as well as sending messages. An electronic bulletin board, to be updated weekly, is planned for late 1994. Electronic mail is regarded as the easiest computerised communication facility for the Central and Eastern European countries to establish (it was felt that this was also true for China). Coverage was still patchy but connections were developing rapidly. Other digital communication links could be difficult to set up between the CCE and Eastern countries such as Ukraine; on the other hand, there were few major problems now with information technology links with Hungary and Poland.

In addition there are informal and formal meetings. Informal contacts take place on ad hoc visits to participating countries and via other academic gatherings such as conferences and seminars. Participants in the CCE network may find themselves collaborating in other international research initiatives. Formal meetings of the Task Force on Mapping, to which the CCE reports, take place annually in Germany. CCE network members also take part in UN-ECE Workshops on acidification and critical loads.

Formal gatherings of the CCE with the national focal centres and their collaborating institutes take place once or twice a year. They take the form of workshops, held in different places around the network member countries in order to promote cohesion, allow researchers to see conditions in a variety of affected environments, and give all countries a sense of 'belonging'. The purpose of the workshops is to bring network members up to date with the current 'state of the art' across the whole field, exchange information and ideas with some researchers invited from outside the network, and discuss issues arising from current and forthcoming needs of the Convention's Task Forces and the UN-ECE policy-makers. Visits are made to sites of scientific interest in the host country.

A potential problem for network members from Central and Eastern Europe is the cost of travel and subsistence for meetings abroad, especially in the West. It was stressed that subsidising travel and subsistence costs for researchers from the new democracies was good practice. An alternative canvassed by one East European scientist was that collaborative international networks should hold as many of their seminars and conferences as possible in the new democracies. This approach, it was argued, would have several benefits:

- travel costs for the Central and East Europeans, and subsidies from the CCE for travel and subsistence, would be reduced;
- more money would be spent in the new democracies on accommodation and other services, thus benefiting the local economy;

- Western researchers would have the opportunity to see many environmental problems which they might otherwise not be able to observe, and which would offer new insights and experience to them.

To date, the CCE network has held one of its workshops in the East – in Katowice, Poland, in 1992.

Communication problems relating to organisational styles and other cultural differences were not regarded as significant, although it was acknowledged that domestic political and economic problems could be disruptive for researchers in parts of Eastern Europe and the CIS countries. It was emphasised that most participants had some experience of networking, whether at national or international level, and that the multi-disciplinary nature of the field tended to condition the scientists to look for collaborative research approaches.

The problems that arose tended to be in relation to the policy 'customers' for the critical loads maps and interpretations. First, not all scientists were equipped to communicate effectively with policy specialists – who might, as one researcher put it, 'be very annoyed with the results' from a political perspective. Policy-makers, he went on, 'sometimes want us to accelerate, sometimes to slow down', and their demands could on occasion be in conflict with the timetable needed to deliver data of the required quality. Where conflicts arose the network was concerned to make sure that the scientific data were of the best possible quality before they were delivered to UN-ECE, even if this meant that some national partners received 'a lot of flak' from their policy customers on occasion. On the whole it was felt that these problems did not disrupt the operation of the network or diminish the quality of work done.

This pointed to the need for the participating institutes to have as CCE respondents said, at least one person to 'front' the research team and act as the liaison point with the CCE and UN-ECE policy-makers. Not all scientists could have the appropriate skills, nor was it necessary that they should: often, it was noted, the scientists least able to communicate well with non-experts were those who were critically important to the research effort.

However, it was felt that the national focal centres and the CCE needed someone with 'multi-functional skills in fund-raising, negotiating and talking to policy people'. These skills were evidently well-developed among a few of the focal centres, and were being fostered elsewhere as researchers gained more experience of the CCE network and other partnerships, and took on new roles in workshops and other events.

Another issue raised was the divergence in analytical approach between the Western and Eastern partners. It was argued by one Polish scientist that the critical loads methodologies and policy for remedial work tended to reflect the problems of pollution encountered in the West. The new democracies had some very localised problems and specific interactions of pollutants and environmental media that ideally required the methodologies and policy prescriptions to be modified.

A British scientist in the network thought that the differences in approach were not a problem: he felt that the Russian methodologies were at times unorthodox but also thought-provoking, and that the end-products were acceptable in quality and in line with those produced elsewhere. He also took the view that the East European institutes were eager to 'blend in completely' with their Western counterparts in terms of methodology. Apart from their severe domestic funding constraints and sometimes backward technology, they were 'indistinguishable' in outlook from the West European researchers.

Outcomes of the network
The CCE network was seen by all those interviewed as an effective exercise in international collaborative science. It was particularly valuable in terms of contacts and collaborations between researchers and institutes, not only within Europe but also taking in North America and Asian countries. There were especially valuable benefits for Central and East European researchers, who now had access to their peers in the West and could learn about new techniques and methodologies, and form new bi- and multilateral partnerships. It was felt that the network was tightly managed and had a high degree of cohesion despite its size: the emphasis on regular deadlines, frequent communication and opportunities for learning was seen as good practice for other scientific collaborative networks. The network benefitted from major economies of scale due to the overall size of the project. Scope economies were less important as these occurred more within countries than across the network as a whole.

Would the research on critical loads be taking place anyway in the absence of the CCE network? It was argued that there was a considerable degree of additionality in the system. Without the CCE framework for research and the commitment of national governments it entailed, research at national level would go on in any case, but its standard would vary widely and methods and data formats would tend to diverge and become less comparable. Dissemination of ideas around the research community would be slower and less even. The researchers in the new democracies of the East would not have as many opportunities to learn new skills and methods, not only concerning critical loads but techniques of more general use, such as Geographic Information Systems software. Finally, the time taken to feed comparable scientific data of high enough quality into the multi-national negotiating systems and policy-making bodies would be much longer. One researcher said that the timetable for delivering 'a usable product' to policy-makers had been 'drastically' reduced as a result of the CCE network.

It was also strongly argued that the existence of the UN- ECE programme did not constrain research outside the collaborative framework or reduce competition. There was, it was noted, no significant funding attached to the CCE network: all of the research funding was allocated by national governments and their agencies. There was extensive dissemination beyond the network, and none of the research partners was putting all of its energies into the critical loads work for the UN-ECE.

The network will soon shift its main focus towards other pollutants (such as nitrogen compounds) and interactions with different environmental media. This will probably require an expansion of partners within the network, and thus new opportunities for spin-off collaborations in other directions. It was expected that the national focal centres would be able to expand their domestic networks of collaborators accordingly; where this was difficult, appropriate support would be sought from ministries to identify suitable candidates.

9 The ESRC's Small Business Programme (Case Study 7)

This is a case study of the Economic and Social Research Council's (ESRC's) Small Business research programme in the UK. It is one of two case studies of a social science rather than a natural science network. As it is an ESRC Programme, it is in many ways more broadly defined and looser than most of the other case studies of networks in this project. However, it is not easy to locate many more tightly specified social science research networks in the UK that span a number of different institutes or universities.

Overview

The Small Business Programme was a four year programme investigating from various angles the role and nature of small and medium-sized enterprises (SMEs) in the UK. Altogether in the programme there were three research centres and in addition 13 separate research projects. There was a co-ordinator of the programme who was specifically excluded from running or participating in any of the particular projects but whose role in the fourth year was to provide an overview of the results of the programme. The total budget of the programme was £1.45m. Of this, £1m was provided by the ESRC and in addition there were four co-sponsors: Barclay's bank, the Rural Development Commission, the European Community and, initially, the Department of Employment, subsequently the Department of Trade and Industry.

The selection of research areas in which an ESRC programme is established is decided through ESRC procedures and committees. The ESRC have a research programmes board and it selects a small number of areas each year for an overall research programme. This approach means that certain areas receive relatively intense funding and research activity at a particular point. However, it also means that researchers who do not participate in the programme are relatively unlikely to be able to obtain ESRC funding in that same area at least during the period of the research programme.

Applications to take part in the research programme had to go through a two stage procedure. In the first stage, a short initial proposal is submitted and reviewed; a number of projects are then invited to proceed and submit a full proposal at the second stage. The choice of projects and research centres is carried out on the basis of peer review. There were no particular problems in the selection procedure. However, a number of those who did not succeed in being awarded research grants were unhappy at the outcome of the selection process. This may reflect in part the fact that the SME area has received increasing amounts of research and policy attention in the last decade and so many new researchers have entered the field.

Motivation

Motivations varied for participating in the research programme. Funding was an important issue but some expressed the view that participating in the research programme and interacting with people in the field was more important. The potential for acquiring new information and for learning from researchers with different methodological strengths was seen as of considerable importance. Initial interests with respect to the research programme varied. Some researchers had little or no conception at the time they applied for their grants as to what a research programme might entail. Others had much clearer expectations of a research programme and saw it as a means of positive interaction with other researchers and of getting up to speed quickly in what for some was a relatively new area. Some perceived there to be additional status obtained from being in an ESRC programme rather than just in having an ESRC grant. This additional status and visibility was seen as particularly important when carrying out research internationally and making international contacts and enquiries. The programme was substantially interdisciplinary – there were researchers from a number of different social science areas including economics, economic history, accountancy, and law. Some participants saw this as a positive motivation for getting involved especially if it would help them to find out relatively quickly and easily about the key approaches in some fields, notably in economics.

Unlike in other networks, where the co-ordinator is often participating in a particular research project, in this case the co-ordinator's role was different. Motivation for taking the co-ordinator's role, therefore, included great interest in the overall programme, the ability to have an influence on the development of the programme, and the possibility of producing a substantial report that would represent the key UK report on SMEs for many years. Interviews were not undertaken with the co-sponsors but their motivations may include access to information and to researchers, and association with the major overall programme and report on SMEs in the UK. There are also status and reputational issues potentially involved.

In many cases, participants already had prior interest in, or plans to work on, specific research topics, and would have looked for alternative funding in the absence of this programme. In other cases, research projects were less likely to have occurred in the absence of this programme and in some cases the nature of the research carried out was different in methodology and approach from that which would have been carried out in the absence of the programme. The motivations for participation and the nature of the programme indicate the main benefits arose due to economies of scope and externalities.

Management and communication

The co-ordinator of the programme had a number of roles. The co-ordinator's role is not identical to that in networks where there are highly interlinked projects working with one specific research goal as their target. One of the functions of the co-ordinator was to bring the various co-sponsors on board with respect to the programme and to liaise with those sponsors. The co-ordinator's role was also to bring the researchers in the programme together on a regular basis through workshops in order to discuss one another's work. Over the time period of the programme there were about ten research workshops altogether. The co-ordinator liaised with the different research projects putting them in touch with relevant people or information and acting as a general co-ordinator and information source. The co-ordinator was also involved in dissemination of the material of the research outputs from the programme and they have run one national conference on the programme and a second is planned. The co-ordinator has also had an additional year in order to write a book about the current state of knowledge on SMEs drawing from the Small Business Programme.

The role of the co-sponsors in the project is a question of interest. The co-sponsors can be seen either as club members in the programme, in which case they are contributing to a central fund and they get benefits from participating in meetings, or they can have a contract relationship which would imply that they were paying for specific projects and had some role in managing those projects. There have been some problems within the management of the programme as three of the co-sponsors were effectively club members but one perceived its role more as that of a contract member. From the researchers' point of view, the researchers had contracts with the ESRC and therefore did not expect to be in a contract research position with respect to co-sponsors.

The main meetings for members of the programme were the workshops held about three times a year. These were one day meetings and about five researchers could present their work at these meetings. These meetings were generally seen as well managed and valuable. They gave researchers the chance to get to know other people in the field, to get to know what research they were doing, what problems and questions they were facing, and to understand the different disciplinary approaches and styles in this field. Benefits were seen to grow at later meetings as participants knew each other better and as the projects developed. There were different views expressed as to the functioning of these meetings in terms of the comments on and reactions to presentations. Some people viewed it as an extremely supportive and helpful atmosphere, while others considered it to be occasionally critical and challenging. These differences in perception may well arise from the different interactive styles of the different disciplines represented. A number of people expressed the view that it would have been beneficial to have longer meetings possibly spread over two days, since this would have given substantially more scope for discussing with other people and getting to know other people.

There was substantial communication between research projects outside of the formal workshop meetings. The amount of this communication varied considerably from project to project. Most of the communication was bilateral rather than any clearer larger groups interacting informally outside of the formal meetings. Many did not know about each other before they entered the programme. Some researchers saw their projects as rather separate from the core focus of the programme and considered they were on the periphery and seen to be on the periphery. Others thought they benefited substantially from the communication flows and that people were more helpful and more responsive than they would have been in the absence of a programme. Even where projects were seen as separate from most other projects, it was found useful to have comments from people who are not directly involved that can bring in different aspects of the subject. This was probably particularly so because of the network's interdisciplinary nature, given that there are other structures where comments from one's own discipline can be obtained.

There was also a question raised as to whether the three research centres were inevitably too dominant in the programme and not necessarily interested in the rest of the programme – this did not seem to be a general view. The centres did find the workshops and the contacts to other people in the network useful, so it was not just a question of the centres doing their own research and ignoring the programme. One centre felt there was some overlap in their research areas with other members of the network which created in effect some creative rivalry and competition and led in the end to a better end product. Overall, links with other researchers were greatly strengthened with a notable lack of competitive tension between the research teams. There was however limited opportunity for collaboration on projects, largely due to lack of time and resources.

Research outputs and outcomes

Within the context of the programme, three books have been produced, co-edited each time by one of the centre directors and the overall programme co-ordinator. In addition, the various projects and research centres have produced many papers in academic journals and working papers produced with a common cover page indicating that they have come out of the Small Business programme.

The programme did not result in one common approach to the study of SMEs and this was not the intention. It did, however, result in some emerging common understanding or consensus on some of the key issues and areas of research and questions to be addressed and a number of stylised facts were seen to have emerged. There is potentially some danger perceived that there could be a narrowing of focus and a loss of diversity as a result but it was not seen to be a great danger.

Although some researchers in the project knew many of the people in the programme, others did not and found it a useful way to make contacts and interact with people. More than one respondent commented that it provided a major resource base to develop information and knowledge in the area significantly different from simply attempting to carry out a literature survey. The co-ordinator had done some very good work in dissemination of papers from conferences, and in suggesting contacts and being a helpful source of information. There was also increased interaction outside of the programme, often because people outside of the programme were aware of its existence and themselves made contact. As a result of the programme, people had given presentations at international conferences, and done work for local economic development agencies in the UK. Some researchers commented that they would have expected the programme to lead to more European and international contacts and were surprised and slightly disappointed that it had not done so.

For those whose research was relatively self-contained and who did not look to obtain much from the programme, the interaction and meetings were seen to be comparable to an ESRC study group and not particularly different from other forms of informal academic interaction. For those whose projects were either more integrated into the programme or who were more aware of benefits they could obtain from the programme, the interaction and meetings provided greater help than would normally be achieved from informal academic interaction. Many welcomed the interdisciplinary nature of the work. However, in one case where interdisciplinary interaction was seen as successful, there were problems back in the researchers' home department and discipline where carrying out interdisciplinary work was not necessarily valued in career structures and career evaluation.

In the interviews carried out, there were no comments that the meetings were totally unproductive even where individual projects felt themselves to be essentially separate from the programme as a whole. However, people did see there being an important trade off as far as time was concerned. Time more than money was seen to be the academics' key constraint. The value of a programme would partly depend on how well networked you were in the particular area already. If participants had the choice between being in a programme and having somewhat more money for their individual research, there was some view that the additional money and the time saving may be preferable. Some people felt that more could have been done with the network and more value could have been got from it through greater interaction and co-operation. The programme was seen as having brought about a step change in the scope and scale of SME research in the UK.

Conclusions

Overall this programme appears to have been very successful. Individual projects and research centres have worked effectively and the programme has operated as a loose network. Opportunities for interaction and collaboration were provided, and the effective co-ordinating management style meant that it was up to individual researchers and projects to take the initiative as to whether they wished to benefit from deeper interaction within the network. Some projects chose to remain only loosely connected within the network, while others chose deeper interaction. This flexibility in participation within the programme seems to be essentially positive. Some difficulties in interaction and communication appear to have arisen due to the interdisciplinary nature of the programme and the perception by some in less well represented disciplines that the more mainstream disciplines were not interested in interdisciplinary work. This is a general problem of interdisciplinary work and not a specific one to this programme. Indeed, despite these comments this programme seems to have provided benefit, due to its interdisciplinary nature, to a number of projects. There was scope for further formal stimulus to the interaction notably through longer meetings but overall the co-ordination of the network was seen as having been extremely successful both overall and in terms of managing bilateral relationships. There may have been a case for allocating funds within the programme to pulling in more international contacts and encouraging participants to develop their international and not only national contacts. The programme provided scope economies through the emphasis on interaction and information exchange, some scale economies in terms of access to a wide range of contacts, and externalities in contributing to future research possibilities due to participation in the network.

10   SME Observatory (Case Study 8)

This is a case study of a European network carrying out a project for the European Commission. It is a social science network and consists of a mixture of institutions – some academic institutes and some more consultancy-oriented institutes. Interviews were undertaken with five different institutes in the network.

Overview

The European Observatory project is designed to produce each year, for Directorate General 23 of the European Commission, a report on small businesses in the 12 EU countries. Its budget is 1.2m ECU per annum. The aim is to produce an overview of the situation and prospects of small and medium sized enterprises (SMEs) that is comparable across countries and that covers a number of different themes on an annual basis. This information can then feed into the European Commission's policy analysis and development. The report that is produced annually for the commission has about eight chapters which deal with specific topics such as entrepreneurialism, finance, and employment. As well as producing an overview of main developments in the SME field, it also has a specific theme each year. In the first year of the project this was internationalisation and its implications for SMEs, in the second year it was the role of the craft sector. The initial contract for the Observatory project is for three years but it is expected that this will be extended or renewed so that the Observatory project will continue after this time. The aim of having a network producing this report is to produce comparable data across countries and to use unpublished or 'grey' literature in the analysis. The main work of the Observatory network is not, therefore, fundamental research but rather the analysis and collation of existing statistics and other information.

The co-ordinators of the network are a large Dutch research centre specialising in SMEs – the EIM Small Business Research and Consultancy. This centre carries out research and consultancy. It had had little previous experience of working collaboratively in a large international network. The co-ordinators had had contact with the Commission prior to the tender for the Observatory project. They had understood that the Commission were dissatisfied with the quality of proposals received for international comparative studies and were keen to identify institutions that could do good quality work on SMEs and that had extensive international contacts. It was also apparent, prior to the publication of the tender document for the Observatory, that this tender would be forthcoming.

The co-ordinators, therefore, decided to establish a network for SME research across the EU both in order to be in a position to tender for the SME Observatory and to be in a strong position to tender for other EU projects. To date, the network as a whole has one additional large project and it has submitted various proposals. A limited number of network partners have worked together on other projects. The network – known as the European Network for SME Research (ENSR) – should be seen as distinct from the specific SME Observatory project. There are 12 organisations in the ENSR which are carrying out the SME Observatory project. It was hoped that all these 12 could be wholly independent organisations and not public agencies but in some countries this proved impossible. Apart from quality of research and ability to do the work, selection criteria also had to take into account which institutes in any country had the highest national profile in order that national representatives within the EU deciding on the tender award would be in favour of this particular network. (Subsequent to the case study, the network was extended to Austria, Finland, Norway and Sweden.)

Many of the organisations involved in the network did not know each other prior to this project. Many of them also did not know the co-ordinators of the project.

There were some problems in the initial establishment of the network largely due to misunderstandings and failures in communication. There were some problems due to the Commission bureaucracy – work started on the first year's report before the contract had been obtained from the European Commission, at the same time some institutes were not in a position to start work without a formal contract. When the co-ordinators established the network, they were involved in substantial amount of time spent on meetings in Brussels, lobbying, and obtaining legal advice. They expected that the other members of the network would contribute to the costs they had incurred, whereas at least some of the other members of the network had understood that they were responsible for their own costs in attending meetings about establishing the network but had not expected to be asked to contribute to the co-ordinators' costs. The co-ordinators are keen that the network should be formally and legally established as a network and have been carrying out discussions with the network partners to attempt to achieve this. It is currently hoped that this will occur and that the network will be registered as a European Economic Interest Grouping. It is anticipated that about 8 or 9 of the current members of the network will be in the formal network. Some of the members of the network will be unable to participate in the legally established network due to the particular situation and constitution of their research institute. Experience with the SME Observatory project has also resulted in some members of the network having some doubts about participating in a legal and formal network.

Motivations

There were various motivations for joining the network and participating in the SME Observatory project. Most participants saw the opportunity to work in an international network as an important opportunity. Funding for the project was an important motivation for some but not for all. The funds were not distributed evenly across members of the network but was proportional to participants' contribution to the Observatory. While some institutes thought they had done fairly well in terms of funding received, others had only received relatively small amounts of money and their contribution was consequently that much less. This was seen as problematic in some but not all cases. The purpose of participating in a project where the contribution and finance was relatively small was in order to be in the network and to use the project and the network as a basis for future collaboration and tenders. Being in the network was seen as allowing institutes to participate in other large projects, so size of project was important.

While some of the participants saw the Observatory project as a way of obtaining additional data and information and also of benefiting from additional expertise in areas where they did not normally focus, others did not see the network as particularly adding to their research expertise or information. At least one member of the network thought that most of the information was actually fairly readily available from alternative sources. Since many of the organisations involved were commercially oriented or operated on a consultancy basis, the research was not seen as inhibiting or altering other research plans rather it was providing funding for contract researchers and enabling larger programmes of work to be developed. For some the work was different to what they would usually do, for example, in doing comparisons of national statistics rather than in doing small sample survey work. There was also seen to be potentially important benefits in terms of status and reputation from participating in this large European-wide SME project. The project was seen as the most prestigious project in Europe so far as SMEs are concerned. Given the nature of the Observatory project, facilitating a larger and wider project than most individual institutes would have undertaken, the network – with respect to this project – is based mainly on economies of scale. Participation in the Observatory project is also hoped to bring externality benefits in terms of other contracts and contacts. Scope economies appear relatively low.

Management and communication

The SME Observatory Project is a rather unusual one in terms of a network since there is one clear product that arises from the network destined for one specific customer. That one product is the annual Observatory report for the Commission. This affects both the management, structure and nature of the work in the project. The network has two or three formal meetings a year for the project with other informal meetings and communication in between. The formal meetings normally last for two days. The formal meetings discuss the work programme for the year and its allocation; later meetings in the year discuss the drafts of chapters and of the report.

The liaison with the Commission is carried out by the co-ordinators who are the main contractors. The Commission does not attend the network meetings. However, other members of the network may attend the co-ordinators' meetings with the Commission if they wish to. Some members of the network have felt that this process was not completely transparent and have felt somewhat excluded from the decision-making process as a result. The network can and has made suggestions and contributions as to what should be in the report each year, in particular what should be the themes of each chapter. These proposals are taken to the Commission by the co-ordinators who then take on board Commission suggestions and wishes and bring this back to the network. Once the topics have been set, the network and individuals are relatively free to decide how to go about them but the Commission comments on the draft chapters.

All the institutes provide information for each of the main chapters. Each chapter has a chapter co-ordinator. This person's job is to collect the required information from the other 11 organisations and to put it together and draft a chapter. This is the most important and interesting task within the Observatory Project and is the main point at which some form of genuine research can be said to be taking place. Where institutes are not chapter co-ordinators, they are essentially in a position of providing information that is requested by the chapter co-ordinators usually to a very tight schedule and with relatively limited time to obtain the information. The extent to which institutes comment on individual chapters has varied and this seems partly to depend on the decisions and behaviour of the individual chapter co-ordinators. The chapter co-ordinator clearly has a fairly powerful role in analysing and interpreting the data for his or her particular chapter.

A number of problems arose in the network due to the question of chapter co-ordination and allocation of these responsibilities. It was agreed that every institute would co-ordinate at least one chapter during the three years of the network. Nevertheless, this means that some institutes including the co-ordinators will be responsible for co-ordinating considerably more than one chapter during that period whereas others will have only one opportunity to co-ordinate a chapter. This has led some of the participating organisations to consider that they are in effect in a sub-contracting relationship to the main co-ordinators and are not in a proper international research network. Other institutes who also had not acted as chapter co-ordinators did not find this problematic. Some researchers saw the information they were required to produce as like responding to a shopping list and found that this was problematic for their own status within their own research institute. There were then important differences of opinion and experience here.

One change in the management and organisation of the network that has been made in response to some of these concerns was that in the second year of the Observatory Project, two consultants on each chapter were appointed to assist the chapter co-ordinator and to advise on analysis and interpretation. This allows for a much wider range of participation across the network. Chapter consultants did not necessarily meet very much – they often communicated by fax and telephone.

Despite these changes some institutes remained concerned about the overall operation and management structure of the network. These concerns seemed to link to a view that they were excluded from important decisions in the network. In response to these concerns, a new steering committee was introduced with three members – from the co-ordinators' institute, and two other institutes. This has a number of roles, including: monitoring output, taking decisions on chapter co-ordinators, allocating budgets and stimulating transparency.

Other concerns raised, again only by some institutes, were that the co-ordinators benefited substantially more from a higher profile through organising the network than did other members of the network. There appeared to be some personality differences and difficulties in the project and unfortunately these had not been resolved through discussion at the start. It was considered by some that the co-ordinators should have used more energy and time in bringing together the partners at the start and in making clear what their expectations should be. This did not appear to be a majority view. It was thought by some that some of the problems and worries arose from lack of transparency and clarity in the operation and management of the network. Measures had been taken to try to improve this for the future. The newsletter was now to appear more frequently to provide network members with information about the project and all members would also receive minutes of the meetings between the co-ordinators and the Commission.

To some considerable extent, the work of the network is essentially a consultancy project. This may underlie some of the problems that have arisen in the operation of the network. Other related problems are also concerned with timing and response to deadlines. Since the research is being carried out for a particular customer with a clear annual product, it is important that deadlines are met and it was considered that not all institutes had a proper appreciation of the importance of this. The co-ordination and research demands were seen as being extremely high. At the same time partner institutes felt they were under considerable pressures in terms of deadlines and budgets. There was a view that differences in approach and experience between the more academic and the more consultancy oriented research institutes led to different responses to the demands of the network.

Another consequence of the structure of the network and the role of the chapter co-ordinators and the need to produce the annual report was that there was relatively little interaction across the network as a whole apart from the periodic partner and expert meetings. Participating institutes were working to supply required information to the relevant chapter co-ordinators, and in the second year some of them were supplying chapter consultants to work with the chapter co-ordinators but in general the work was not that interactive since the main task was to provide work on each institute's own country.The extent to which it is perceived as an interactive network clearly depends on whether an institute is a chapter co-ordinator or not. Nevertheless, contacts were closer than they would have been if people were just meeting at conferences.

There was seen to be very few strong bilateral relations that had developed between any of the partners. This was seen as reflecting the fact or the agreement that the network as a whole had to be involved in everything and that the co-ordinators were keen to be at the centre of the network. Whereas for some institutes the network will clearly form an important basis for contacts for future European work, for others they already have a wide range of European contacts and will not necessarily go to the network for all their future partners.

Communication other than through meetings was by telephone, fax and to a lesser extent e-mail. There were no major perceived problems in terms of communication at this level.

Outputs

The main direct output from the Observatory Project is simply the annual report for the Commission. All the institutes are acknowledged in this document. The extent of other benefits from participating in the network appeared to vary across institutes – some felt it had given them considerable access to other contacts and was of considerable use in other pieces of work. There was a view that many more collaborative projects and tenders could have arisen from the network than actually had arisen so far. Nevertheless, most members of the network still do hope that this will provide a basis for many future projects.

Some have used the experience of network building to extend their contacts within and beyond the EU and even to develop links with non-research bodies, eg. with chambers of commerce across Europe. The Observatory Project is seen as an excellent marketing tool for all partners – this is seen as a key benefit. It has also brought access to many contacts and major sources of information within and beyond the EU. For some academic groups they were used to being in networks mostly with universities, so this network had brought them in touch with a different type of institute and researcher. Another outcome is press and media coverage. There is a lot of press coverage associated with the report and there are other contacts also, for example, presentations for members of the European Parliament.

Members of the network can use the information in the network to produce other papers and articles. There did not appear to have been any clear agreement or discussion of whether all the institutes had equal access and freedom to use all the information that the network was producing and whether there could potentially be conflicts of two different institutes producing similar papers or articles. If the data being produced is accessible from other sources – the view of one institute though not of most – this would clearly be less of a problem. There were not seen to be any major differences between a national and international project. The programme had encouraged synergy, common understanding and consensus which was seen as entirely positive. The difference from working in a single institute is the nature of the communication and the set-up costs.

Conclusions

This is an example of a European research network established in the first instance to work on the SME Observatory project that is as much a consultancy project as a research project. Many members of the project saw it as very valuable not only or even mostly in itself but in allowing the network to be established and to raise their profile in European circles. However, some members of the network were less happy with their experience and this seems to result partly from the specific nature of the project but also from management problems in the initial establishment and running of the network. The lack of experience of managing a large international network of the co-ordinators, together with time pressures, may have played a role here, nevertheless, it seems that the network has acted to improve communication and managerial questions and there has clearly been an important learning process here. Main benefits are: funding, participation in a large project, contacts, international profile, and the opportunity to undertake future projects on the basis of the existing project. Economies of scope are of some importance in drawing on the different expertise, especially country-specific knowledge, of the different institutes. However, the main basis for, and benefits from, the network are economies of scale. There is the potential for more interaction than has occurred. This project is slightly unusual in that the potential for future benefits from being in a network appears to be as important a motivation as undertaking the project i.e. an externality benefit as much as a scale benefit.

11 Overview and Conclusions

This chapter sets out a comparative analysis of the eight case studies of cooperative research networks carried out in the study. In order to facilitate reference to the case studies in this chapter, acronyms will be used for each of the case studies. They are as follows:

Case Study 1   Modelling turbulence – RR
Case Study 2   European computational aerodynamics research programme – ECARP
Case Study 3   The animal cell bio-technology club – ACBC
Case Study 4   Peroxidases in agriculture and the environment – PAE
Case Study 5   Critical loads advisory group – CLAG
Case Study 6   Mapping critical loads for the UN-ECE – CCE
Case Study 7   The ESRC's Small Business programme – SME
Case Study 8   SME Observatory – OBS

As discussed above, in analysing co-operative research networks, it is possible to identify three main types of network and combinations of these three main types. The three main types are as follows:
(i)   Separate but related research projects
(ii) Interdependent projects/joint research
(iii) Individual projects whose joint or collective output also constitutes a project – and the overall project imposes constraints on the individual projects.

The classification of the eight studies according to these types or to combinations of these types of networks is set out in table 1.

As is clear from table 1, most of the networks – five out of the eight – are either full or hybrid type (ii) networks and involve substantial amounts of joint research. Two of the networks have been classified to type (i) – ACBC and SME. Although one of these is a science and one a social science discipline they have in common that they were deliberately constituted as rather broader programmes than the many more tightly specified research networks. One was a SERC club and the other was an ESRC programme. Only one of the eight case studies is classified to the type (iii) category – OBS. It is only in this third category that networks may be based purely on scale economies. The other networks, given their classification, will depend on a combination of scale and scope economies although those in the pure type (i) category may be unlikely to benefit substantially from scale economies. All of the case studies may benefit from externalities in any of the three main categories. The relative importance and nature of scale and scope economies and externalities is discussed further below. It is clear that the eight case studies represent a range of types of networks. They, therefore, provide a good basis for comparison of network experience.

Table 1	Network types		
Type	(i)	ACBC	SME
Hybrid	(i) / (ii)	RR	
Type	(ii)	PAE	CLAG
Hybrid	(ii) / (iii)	ECARP	CCE
Type	(iii)	OBS	

While there is no particular pattern apparent by subject area in table 1, there are apparent links between network types and whether networks are national or international. Three of the European networks are related to type (iii) in their categorisation while only the national networks have type (i) classifications. This is largely a question of size, which is discussed further below. It may also indicate that looser networks are more appropriate to a national context but this is speculative.

Motivations

Participants have a wide variety of motivations for joining, and carrying out research in, a network. However, a number of common, key motivations do emerge across the case studies. Funding is a prime motivation for most participants in networks. However, it is not the key motivation for all participants. Some of the interviewees in the case studies were not motivated by funding either because they had substantial funding possibilities elsewhere or because they were actually receiving extremely low funding in their network. They, therefore, had other motivations. In some cases, these other motivations were still financial in that participants were hoping for future funding from existing participation. However, in other cases, funding was not a key or important motivation and in these cases motivations such as contacts and sharing of research expertise were the prime motivations.

Virtually all scientists interviewed had motivations other than funding from participating in networks. Other key motivations included: participating in a larger research project than would otherwise be possible; participating in a different research project than would otherwise be feasible due to differences in expertise across members of the network; acquiring new expertise, entering a new area; obtaining up-to-date information and scientific results; widening contacts; status; and opportunities for further collaboration and fund-raising. It is clear from these motivations that scientists are participating in networks for a range of reasons that can be categorised as either economies of scale, economies of scope or externalities. In those networks where there was either industrial participation or policy-makers participating, this was in general seen as an additional positive motivation for joining. However, in the case of the CLAG network, there were some differences of opinion about the costs and benefits of researching on a project in such close and interactive proximity with policy-makers.

The balance or relative importance of funding versus other motivations is an important policy consideration. If scientists were participating in networks purely in order to obtain funds, this would suggest that network structures for research are actually inappropriate. Even if policy-makers wish to direct research in particular directions through the allocation of funds, this is still not a justification for directing those funds to research in networks. However, EU policy makers may focus, even in these circumstances, on funding for networks for two reasons: firstly, in order to encourage cross-European interaction and understanding, secondly, they may consider that if they do not fund certain areas, the nation states themselves will not fund those areas, and that if they are to justify EU funding, it should be provided for network structures. Thus, the Commission may stress collaboration even where it is not the best solution in order to justify their involvement in terms of subsidiarity (discussed further above – chapter one). There may therefore be a variety of different influences on policy-makers. Since participants in networks did have other motivations for being within the networks, the problem is not so great as if funding were the sole motivation. Nevertheless, it is important for policy-makers to assess the weight of these other motivations relative to funding if the appropriateness of network forms of research is to be evaluated. As was discussed above, alternatives to network organisation include bringing researchers together into a single research facility and providing for more informal means of interaction through standard scientific and academic channels such as conferences, seminars and societies. Alternative funding structures may also be necessary for smaller, collaborative groups – bilateral or trilateral.

Research focus

In most of the case studies, scientists did not think that they were being particularly forced into different research areas in order to obtain funding. Where research direction was affected this was largely seen as positive. However, effects on overall research focus within and across disciplines cannot be assessed only on the basis of the views of those who received funding – the results presented here therefore only address one aspect of this issue.

In most cases, scientists considered they would probably have been working in the same or a similar area, though not an identical project, even if they had not participated in the network – a similar finding to the UK IMPACT study (1993).

They would have looked for and expected to find some alternative funding elsewhere. This was most strongly the case in two networks (PAE, SME). In other cases, research area and direction were affected by the purpose and aims of the network but this was in general not seen as problematic rather it was seen as a normal way for research to proceed. Thus, in the case of the two aerospace networks, scientists were accustomed to working with industry and positively welcomed the opportunity for close research relations with firms in the field. It was important for their own scientific development to understand the problems the firms were facing. Similarly, in the area of environmental science, researchers recognised in most cases that their research did have substantial policy relevance and most of the researchers, but not all, welcomed the opportunity to interact closely with policy-makers. In the case of the two looser networks – ACBC and SME – both of these networks were trying to either establish or increase substantially research in their area of interest. This was strongest in the ACBC case: scientists were responding to decisions by policy-makers and research managers to move into this area. In the absence of funding, much less research in that area would have been undertaken. Similarly in the case of the SME network, while a number of researchers would undoubtedly have carried on with research in this area, the programme enabled a substantial increase in the scale and scope of UK SME research to take place. In the case of the European SME project – OBS – participants were undertaking work that they may not have done otherwise even though it was closely linked to their existing areas of expertise. This is linked to the fact that, of the eight case studies, this one is the closest to a consultancy project as much as to a research project.

There is a balance to be made in evaluating the effects on research of participating in a network. On the one hand, the question arises as to whether scientists' research is being distorted in ways the scientists would not choose if they were not subject to funding constraints. On the other hand, policy-makers may be concerned, especially at EU level, to ensure that they are funding research that would not have occurred otherwise so that there is additionality. Research additionality – for example, undertaking a large project that no one country could have done – is not the only justification for networks or for EU action. The main justification is that networks and EU-level projects should be more efficient than alternative structures.

Management and communication

The role of management and co-ordination in establishing a network and in affecting its successful operation is a critical one. The nature of the managerial role can make a crucial difference to the effective and successful functioning of the network. There is however no one particular style of management appropriate to managing a network. Many factors will influence the actual and desirable type of management. These factors will include: the aims of the network, the scientific area, the skills and contributions of the different participants, the size of the network, funding, and length of time the network will be in operation. The structure and operation of the network will depend on these factors as well as on management decisions.

Successful management is not equivalent to tight managerial control. Some networks will need to be more tightly managed than others where specific joint or common outputs are required, especially where there is a need for common formats for outputs and/or for methods of working. Other networks can be much more loosely and flexibly managed. Since a network is carrying out a range of activities with a range of possible outputs and obligations, not all activities in the network will need to be managed in the same way. Thus, some aspects of a network's operation may need tight management but at the same time it may be feasible and desirable to have a relative loose managerial style with respect to other aspects of a project. One example of this mixture of management styles within the same network is the Rolls Royce case study. Management in this case study can be described as tight in the sense that the overall strategic aims and specific projects were clearly set out and it can also be described as loose in the sense that structures for reporting and for interaction were not completely formal, fixed or centralised.

Given these differences in management within a network, it is not possible to allocate a network to a type of management precisely but it is nevertheless possible to make a general overall assessment of the tightness or otherwise of the managerial control in the case study networks. Thus, for example, in the case of Rolls Royce, the network can be said to have medium strength of management given its combination of tight and loose managerial patterns. Categorisation of the case studies on this basis is set out in table 2.

Table 2   Categorisation by overall managerial approach

Tight	Medium	Loose
OBS	RR	PAE
ACBC	CCE	SME
	CLAG	
	ECARP	

As can be seen from table 2, only two of the case studies are categorised in the 'tight' management category. The remaining six are either in 'medium' or 'loose'. The two case studies classified in the 'tight' management category – OBS and ACBC – are very different types of network. OBS had one clear goal and joint output to achieve whereas ACBC was trying to establish a new area of science within the UK. Similarly the two categorised to the 'loose' category are also rather different types of network. PAE is a highly interdependent project with a clear set of common goals whereas the SME network is a much looser broader structure. The nature of managerial control does not, therefore, appear to be directly related either to the specific aims of the network nor to its overall size.

One key managerial task is to ensure that the network focuses on and achieves, as far as possible, its main aims. This may involve simply or mainly a monitoring role or it may involve a much more active management role. This will depend in part on the tightness and clarity of the goals. But even where there are clear, specific goals management may still only have to play a monitoring role – depending on the nature of the science and the nature of the network. Where management has to play a more active role, is usually where researchers and institutes within a network are required to follow similar or common frameworks or to operate in certain standard ways. In addition, where results need to be brought together across a range of groups and collated and/or analysed together at one point then there would be need for a more active management role. The three case studies where this is most clearly the case are: ECARP, CCE, and OBS. These are all networks where there are some economies of scale to be obtained due to the overall size of the network and the joint nature of the project.

It is clear that there are substantial learning effects involved in networks, in particular related to management, both from the managerial and from the participants' points of view and related to communication and interaction more generally. Networks set-up for a very short period of time, such as a year, may then be inappropriate. None of the case studies operated for such a short length of time, but in the case of the ECARP network there was a view that two years was an insufficiently long period of time for the benefits of learning and operating in such a large scale network to be realised. Some of the networks – in particular RR and CLAG – benefited in their learning processes by being relatively flexible and evolutionary in structure over time. While the clarity of aims and structure appears to be an important aspect of successful management, this clarity should not be achieved at the expense of flexibility. In addition to flexibility, researchers and scientists in general need to be working in a situation where they have a substantial degree of autonomy. Even where clear output requirements, deadlines and so forth exist, it should be possible for a network to be managed in this way.

In three of the eight case studies, there were some problems and/or difficulties with respect to management according to the views of participants. These three were ECARP, ACBC, and OBS. In the case of ECARP, the managerial problems appeared to arise essentially from the difficulties of managing an exceptionally large network – 36 institutes. This put considerable pressure on the managerial skills of the main co-ordinator and sub co-ordinators and their prior experience in networks was critical to being able to manage such a large network. The large network size meant formal structures had to be imposed but at the same time it was difficult to ensure that all of the large numbers of institutes followed those structures. From the participants' point of view, there were also difficulties or weaknesses as a result of being part of such a large structure. The managerial difficulties in the case of the ACBC study are complex but appear to be related to tight managerial control and to variation over time, and differences of opinions, as to the role and nature of the steering group that was responsible for managing the club. In the case of the Observatory – OBS – managerial problems appeared to arise in part from initial lack of transparency and also in part appeared to be related to the consultancy nature of the work and the tight management that was therefore required.

It is possible very loosely to categorise the case studies according to the overall perceived benefits of being in the network – that is the nature of scale, scope and externality benefits participants thought had been achieved. Perceived benefits can be grouped according to whether they are high, medium or low. If this is done and these perceived benefits are related to the nature of the managerial approach some relationship does appear. This is set out in table 3. It is apparent from table 3 that there appears to be some evidence of a negative relationship between the tightness of managerial control and the perceived benefits of being in a network. This suggests that scientists and researchers operate best where they have reasonable levels of flexibility and autonomy.

Table 3	The relationship between managerial control and perceived benefits of network participation		
	Perceived benefits of network participation		
Overall managerial approach	High	Medium	Low
Tight		OBS	ACBC
Medium	RR CCE CLAG	ECARP	
Loose	PAE	SME	

There is no clear relationship between perceived benefits and the clarity of the link within a network to industrialists or policy-makers. This is not unexpected: some research can benefit strongly from such links, but other research is equally or more likely to be important and successful without such links. Nor is it the case that where there are strong industry and/or policy links that management must be extremely tight. In such cases, there probably needs to be a medium degree of control, but tight control is not necessary or desirable.

Interaction and communication
The amount of interaction within a network is one of the major sources of benefits from operating in a network and indeed one of the main justifications for being in a network. Table 4 shows the relationship between the managerial approach and the amount of interaction within a network on the basis of these case studies. It shows that there is a negative relationship between the overall strength or

degree of managerial control and the extent of the interaction within a network. This suggests that, within different managerial approaches, care must be taken to ensure that interaction is allowed and encouraged. Such interaction has to some extent to be informal and flexible. Interaction cannot be tightly controlled and managed from the centre, otherwise there is only an interaction between the centre and the constituent parts. If different groups within a network are to interact with each other and not simply through the centre then there inevitably has to be some loosening of overall managerial control.

Table 4	The relationship between managerial control and extent of interaction in a network		
	Extent of interaction		
Overall managerial approach	High	Medium	Low
Tight			ACBC OBS
Medium	RR CLAG	ECARP CCE	
Loose	PAE	SME	

While there were few, recognised communication problems across groups, there were cases where interaction was less than had been hoped for. Scope for more interaction and so more value-added from the networks was identified in the case of the ECARP, ACBC and OBS studies. In all these cases, it appeared to be due to a combination of funding level, nature of the project and managerial difficulties. The potential for more interaction was acknowledged by management but also recognised by at least some scientists. In the ECARP and ACBC cases, it was seen as managerially difficult to do anything additional to stimulate further interaction. In the case of ECARP, the problem appeared to be to do with the combination of the large size of the network and the limited funding given the overall size of the network. In the ACBC case, the reasons were less clear but projects were fairly distinct within this network and it was also a network where many scientists were entering a new area. Hence, some scientists had substantially more learning to go through than others. This may not have facilitated interaction and exchange of expertise. In the OBS case, funding levels and nature of the project were important factors; however, action had been taken to change network structures and so to increase interaction.

A central function of management in all networks is, therefore, to facilitate interaction, communication and information flows. This is not only the responsibility of management. Communication and interaction depend also on: the initiative and activity of the researchers; the nature of the science and of the

Table 5   Interaction and perceived benefits of network participation

	Perceived benefits of network		
Interaction	High	Medium	Low
High	PAE CLAG RR		
Medium	CCE	ECARP SME	
Low		OBS	ACBC

particular project; the funding levels of the network; and the ability to travel and communicate easily. In the eight case studies, there appears to be a strong positive relation between the extent of interaction and the perceived benefits of being in a network. This relationship is indicated in table 5 through an approximate categorisation of the case studies to benefit and interaction categories. It is relevant to ask what other factors influence or are related to interaction, and in particular the relationship between interaction and availability of scale and scope economies. This is discussed further below.

Clarity and transparency
Management has various roles in facilitating communication and interaction. Management can act as a central node for collecting and transmitting information relevant to the different groups in the network. Management can and should act to ensure that the operation of the network is as transparent as possible and that groups have adequate information on network organisation, structure and management demands. Furthermore, management is responsible for organising the main meetings and gatherings of the network. This will include meetings of the overall network, but it will also usually include meetings of sub-groups of the network. Meetings of sub-groups may in some cases be the responsibility of co-ordinators within the network or it may be left to the initiative of individual groups and researchers.

In all the networks, meetings organised by the network co-ordinators or managers were, in general, seen as valuable and welcome. They provided an important means of obtaining an overview of work within the network. They also provided an important means of getting to know the different researchers within the network and of exchanging information, ideas, expertise, and knowledge. This exchange of information and knowledge was not only limited to work within the network but extended to other areas and possible projects. There were then substantial externality benefits to be gained from interaction and meetings within the network. Many of the networks also found it valuable to ensure that younger or more junior researchers from within groups did attend network meetings and that it was not only the senior representatives from groups who benefitted from meetings and interaction. In more than one case study, the view was expressed that more benefit could have been obtained from meetings if they had been set up for a longer period of time. There is inevitably a trade-off between benefits of taking a longer time to meet and get to know people and to hear about research within a project relative to the pressures on most researchers' time leading to a demand for relatively brief meetings.

Clarity in the management role is also very important. Transparency of information in the network has already been mentioned, and it is important that the reporting and administrative demands of the network on groups should be made clear as far as possible from the start of the network. It is also important that there is clarity at the start on the expected role of management and what aspects the network they will or will not monitor or control. Management may need flexibility and so the precise details of its role do not necessarily need to be classified; management and network structures may need to change over time. Allowing for flexibility and evolution appears to be positive in its impact on network operation. However, allowing for flexibility and evolution does not mean that there needs to be lack of clarity in the initial role of management. Clarity in the management's initial role means that there should then also be clarity and discussion about any subsequent changes in that role. Where management's role is not clear and where there are changes over time without agreement, problems can be caused as in the case of the ACBC study.

Information and knowledge exchange
There is clearly a learning process that goes on over time within a network and network meetings are an important forum where learning and trust develop. In all the networks, the view was expressed by at least some participants that information exchange, knowledge transfer and transfer of expertise were more intense and easier to obtain within the network than through more informal academic interaction. This seemed to be related both to the process of learning and build-up of trust and also to the fact that people were engaged on a common project and had some common interests and obligations including reciprocity in information and knowledge exchange. Interestingly, this greater effectiveness of networks in the process of transfer of knowledge and expertise also extended to obtaining relevant, up-to-date information on a scientific area or field. More relevant and recent information was seen to be obtained in this way than through other informal methods of academic interaction such as conferences, seminars, workshops and requests for working papers. This appeared partly to be a result of the more focused and committed nature of a network relative to a conference or seminar series, but it was also related to the more flexible forms of discussion and communication that take place within a network relative to one of presentation over a paper at a seminar or a conference. This information benefit appeared to hold even where networks were relatively loose and projects were

separate though overlapping such as the SME network. At the same time, scientists were alert to the fact that some of their 'competitive advantage' came from precise details of a process and would not necessarily share this information.

The positive effects of networks on information and knowledge exchange relative to more informal structures are striking. A similar process may be occurring to that identified by von Hippel (1987) of informal know-how trading between firms. In von Hippel's study, the importance of reciprocity and repeat interaction were clear. Networks may act to give scientists more formal structures to operate within, where the existence of some collective commitment and repeat interaction may facilitate trust and learning and so increase knowledge exchange. This suggests potentially major benefits from networks not identified in the UK IMPACT study (1993) which expressed concern that skill effects were only at a junior level.

Communication problems
There were relatively few communication problems in the case study networks. The communication problems that did exist were essentially problems of communication between management and groups in the networks and not so much communication problems across groups. In one case of communication problems between groups, this was connected to perceived lack of commitment, or slowness of response, by one group due in part to its involvement in a large number of networks. The problems of excessive managerial commitments were highlighted in an earlier study (Franklin, 1988). Further problems seemed to arise where there was a view that work within a network was not being shared out fairly and those who thought they had received inadequate funding or inadequate parts of a joint research topic felt resentment towards other groups. Apart from such cases, communication difficulties, if they existed, were seen as largely due to questions of personality and also likely to be resolved through learning. Communication difficulties were not particularly found with respect to: geographical distance, cultural difference, language difference, or other national differences. Personality and perhaps skill differences were more likely to cause any difficulties. Geographical distance within Europe was not seen as a major factor affecting interaction and communication, although distances to the USA and Japan were still of substantial importance. One participant expressed the view that unless you were working together in the same laboratory, it made relatively little difference whether your collaborators were three hundred metres or three hundred miles away. While this may be an extreme view, in general distance was not viewed as an important factor.

Less favoured regions and knowledge transfer
Skill and knowledge differences between less favoured regions in the EU and the more favoured regions were not seen as highly problematic. It was well recognised by most that the EU had more than one motivation in funding and encouraging collaborative co-operative research networks and that one of these

motivations was to aid in knowledge transfer and skills development in the less favoured regions. In general, this was recognised as an additional demand or requirement of being in a network and was mostly not seen as a negative factor. Nor was it necessarily the case that knowledge transfer was all one way; in most cases partners in less favoured regions were in a position to transfer knowledge back to other groups. However, it can be the case that groups from less favoured regions are included in network applications even if the science is not actually appropriate to the particular network in order to promote the network applications chance of success. In one case study, one participant had the opinion that this had been the case. Overall, however, this was not seen as problematic.

In general, participants either saw very little difference across countries as science was international or they saw international collaboration as highly beneficial due to diversity of knowledge and approach. Even where the main science and knowledge base was extremely similar across groups, different institutes and different countries might have slightly different approaches or emphases and these minor differences could be of quite substantial importance.

Means of communication

The extent of informal communication outside the formal structures of the networks varied across the case studies but in all cases there was some. This informal communication took the form in some cases of additional meetings but mostly it took the form of telephone, fax, and e-mail. Fax was in general seen as the most important means of communication. While some networks were using e-mail and found it of substantial benefit, this was a minority of the case studies. In particular, in those case studies where industry was involved, e-mail was less likely to be used due to industry concerns about confidentiality. E-mail use appeared likely to increase from the comments of a number of participants and has the capacity to reduce the managerial task substantially even relative to the use of fax. More direct encouragement to the use of e-mail could usefully be given to networks and to managers of networks at the start of projects.

Economies of scale, economies of scope and externalities.

In general, networks obtain benefits from a mixture of economies of scale, economies of scope and externalities. It is possible, for the eight studies undertaken, to identify the main type of network, as discussed above. The case studies allow an assessment to be made of the relative importance of economies of scale and scope and externalities. Most if not all of the case studies benefit from some degree of economies of scope. Some of them benefit from scale economies and all of them benefit to some extent from externalities. None of the networks seemed to be based purely or primarily on externalities and if this was the case it would have to be questioned whether the network form of organisation was appropriate relative to conferences, societies and so forth. Nevertheless, as described earlier, externality benefits in terms of information exchange can in fact be an important part of being in a network. Whether such information

Table 6   Scale economies (from equipment, samples, data) and interaction

Scale economies via equipment, samples, data	Interaction		
	High	Medium	Low
High	PAE		
Medium	RR CLAG	ECARP	ACBC
Low		CCE SME	OBS

benefits would exist in networks based purely on externalities is open to question and to further research.

Types of scale economies
In considering economies of scale within networks and their influence on interaction and the overall value and perceived benefits of a network, it is helpful and appropriate to identify different types of scale economies. Two in particular can be identified. Scale economies may be achieved due to fixed costs associated with shared use of a particular piece of equipment or shared use of certain samples or data. The second type of scale economies comes from the benefits occurring due to the overall size of a project. Thus, for example, there are scale economies based on equipment in the PAE network but not in particular due to overall size, while in the OBS network there are scale economies essentially due to the overall size of the network. It may be expected that there will be more interaction where there is sharing across a network due to fixed costs rather than where there is scale simply due to size. This is indeed the case. There is no obvious relationship comparing across the eight case studies, between interaction and scale due to size but there is some weak positive relationship between interaction and scale by equipment. This is set out in table 6. If the two different aspects of scale economies are combined to give an overall assessment of the importance of scale economies in the network, there is no clear relationship between interaction and scale economies. This suggests that research which aims to identify the costs and benefits of networks with respect to scale economies should identify the nature of the scale economy and not just take scale as an overall category.

Scope economies, interaction and diversity
There is a positive relationship between the extent of scope economies, including knowledge transfer and interdisciplinarity, and interaction. Approximate categorisations of the case studies are set out in table 7. Apart from the Rolls Royce case study which had high interaction with medium scope economies,

Table 7   Scope economies and interaction

	Interaction		
Scope economies	High	Medium	Low
High	PAE CLAG		
Medium	RR	ECARP CCE SME	
Low			OBS SME

there is a very clear positive relationship between interaction and scope economies – with high interaction and high scope economies for two case studies, three in the medium interaction/medium scope category and two in low interaction/low scope.

In general, the extent of interaction in a network can be seen to be an important aspect of its successful functioning and this interaction comes clearly from scope economies but also to some extent from scale economies. Since the scale economies factor in equipment, samples and data is of some importance in promoting interaction but scale via overall size is not then one lesson is that interaction derives substantially from diversity across groups. Diversity is also likely to underpin a number of, though not all, externality benefits.

Networks appear, therefore, to be particularly beneficial where there is considerable diversity among participants. This requires however the aims of the network to be clearly specified and that it is feasible and beneficial for scientists with diverse skills or approaches to interact and work together. Even where networks are not strongly interdisciplinary, case studies indicated that substantial benefits were gained purely from the differences in approaches that scientists in the same area may take in different institutes or in different countries. These differences are not large – it is clear that science is international – nevertheless small differences in focus, emphasis or approach can be very valuable and lead to beneficial information exchange in joint research. In this context, Europe is seen as having substantially greater strength than the Americans or Japanese. European diversity, rather than being an impediment to the achievement of economies of scale, is viewed as a key factor in providing positive economies of scope that can be exploited within co-operative networks. Overall, the case studies indicate a complex set of relationships between scale and scope economies, interaction, managerial control and perceived benefits. On the basis of these case studies, the more successful networks are those with high scope economies, high interaction and medium to loose managerial control.

While this analysis suggests that interaction can be facilitated and encouraged and that this depends on management, funding levels, network aims, structure and operation, and participants, there are limits to this encouragement of interaction. Interaction also depends on the characteristics of the project, which will include the characteristics of the science, the characteristics of the participants, and more generally the nature and extent of the scale and scope economies that exist. The policy implication, therefore, is not that networks should be encouraged to stimulate interaction artificially because interaction is seen as an indicator of success of a network. Rather, the policy implication is that networks should be assessed especially in terms of the likelihood and basis of any interaction. Where it appears that interaction is likely to be low, a strong case needs to be made with respect to anticipated benefits and potential value added from the network. Where interaction in the network ex post is less than anticipated ex ante, the reasons for this should also be evaluated; it may provide an indicator of problems in the functioning of the network.

With respect to diversity, the policy implication is not and should not be that diversity per se in network participants should be encouraged. Diversity can be critical in leading to knowledge transfer and interaction, but excessive diversity could equally lead to lack of inter-related interests and knowledge and so to less contact. Diversity is important but it needs to be assessed in the context of the aims of the project, and the nature of the science to see how it contributes to scale or scope economies.

Outputs and additionality

The overall benefits and effects of network operation if there are positive economies of scale and scope are that research may be carried out more quickly, at lower cost and more effectively than otherwise, and that research is carried out that would not or could not be carried out otherwise. The main direct output from networks is usually in terms of papers and books but other outcomes, mostly externalities, also exist, such as the establishment of national or international societies, the establishment of a national or European reputation in an area and the possibility of future funding, together with the on-going benefits from knowledge transfer, information exchange, and so forth. Additionality – in the sense of whether the project would have been carried out otherwise – is not necessarily a good test of the appropriateness or success of a network. Networks may cause researchers to focus on an area or project that they would not have chosen otherwise and this may be beneficial as long as the motivation is not purely financial. However, the fact that research may have been carried out anyway is not an indication that a network should not have existed. The test is whether there are synergy benefits from carrying out the research within a network rather than in other organisational forms, i.e. whether there are indeed economies of scale and scope from network operation.

National and international networks

Participants in networks do not perceive any substantial differences between national and international networks, especially in participating in European relative to domestic networks. Geographical distance in Europe was seen as relatively unimportant (though not to the US and Japan). Cultural differences were seen as relatively unimportant and language differences were also not perceived as problematic. Difficulties in network communication – apart from managerial problems – were seen as largely due to personality differences and skill differences rather than to cultural, national or geographical differences. The main reason for this lack of perceived difference is due to the international nature of science and the increasing familiarity of scientists with collaborating internationally. At the same time, as discussed above, participants in networks did see benefits arising from diversity and this diversity was often due to relatively small differences of approach or specialisation which may be partly linked to different emphasis in science institutions and science policies in different countries. Diversity is then not perceived as problematic or as causing major differences between national and international networks but at the same time there are positive aspects to diversity which in fact add to the benefits of network operation. Motivations for participating in networks were similar for participants from different countries. For scientists in less favoured regions, it was often the case that there were particular benefits perceived from working with some of the leading groups in Europe who were also in the network. Questions concerning management and communication issues within the networks and the potential problems that can arise do not appear to vary across national and international networks – other than that international networks may be larger.

Although there was a lack of perceived differences between national and international networks, a comparison of the case studies does suggest some differences. In particular, as was illustrated in table 1 above, there are differences in the eight case studies in the type of networks that are national and international. Both national and international networks may be based on joint research and strong interdependencies – type (ii) networks. However, where large scale economies exist due to benefits of size and of a larger project, rather than through equipment, there appears to be a much greater likelihood that these will be European rather than national networks. All the four European networks were classified to either the type (ii) or type (iii) networks, while all the UK networks were classified to either type (i) or (ii). Type (i) networks are those where researchers are primarily working on separate research projects but where these may in various ways overlap and so there are benefits to be gained from interacting and communicating across projects. These tend to be looser networks and the benefits do arise from a lot of informal as well as formal interaction and communication. It may then not only be chance that from these eight case studies it is only three UK ones that fit into either the type (i) network or a hybrid (i)/(ii) network. Where looser and more informal networks are concerned it may be the case that national networks have some advantage. This would need to be investigated through further research.

Comparisons by subject and discipline

While many of the issues considered appear to be common across networks, particularly questions to do with management and communication, there are some differences across subjects and also between natural science case studies and social science case studies.

Externalities, however, appear to be similar across subjects: information, knowledge, contacts, future funding, and status. Scope economies are likely to be especially important where networks and projects are interdisciplinary and for three of the four interdisciplinary case studies this was the case: PAE, CCE, CLAG. It was only in the case of the ACBC case study that, though there was interdisciplinary research, there was not much interaction and so not much benefit gained from economies of scope. However, there was the potential for minor scope economies in all the networks, and although these may be minor in terms of the nature of the difference between groups and participants, they may be important in terms of the cumulative benefits of knowledge and expertise and information transfer. Benefits from scale economies where the scale economies are based on shared equipment and samples are potentially relevant to all areas. Where large size of the network and the project is important this may perhaps be more likely within a single subject – for example, ECARP and OBS. The environmental science case study – CCE – is an example of an interdisciplinary network where large size is important, but most of the interdisciplinary interaction was within the networks within each country, whereas the interaction across countries was actually based on similarity of approach and knowledge, the differences being that each national network knew about each national situation.

This question of national expertise is relevant in a number of cases. It is not a question of whether science is national or international, but whether the particular subject of enquiry is national even though internationally accepted methods of research are being used. Thus in the case studies in environmental science and in economics, the participants in the networks were mostly focusing on their own countries whereas in the bio-technology and aerospace areas this was not the case. This would imply that the nature of the interaction and exchange is different in networks where there is and is not a national focus. This is likely to affect the extent and nature of interaction within a network.

With respect to the social science pair of case studies relative to the natural science case studies, a comparison can only suggest some possible differences given the small number of social science – economics – case studies. In general, it may be expected that there will be weak or no scale economies in equipment in social science case studies though there may be scale economies with respect to data sources. There may be scale economies through carrying out a larger project and this was the basis of the OBS project. There are some scope economies or there can be scope economies in social science networks, especially but not only where these are interdisciplinary or where there is specific country or individual expertise. There were more scope economies in the national network – SME – due to the fact that it was an interdisciplinary network unlike OBS. In general, it may be expected that scope economies would be more important for a social science network than scale economies but at the same time social science in general may be less interdisciplinary and less based on diversity than natural science and for this reason there would be less motivation for large-scale joint research. Nevertheless, there are benefits to be gained on the basis of scope economies, just as there are in the natural sciences, even where networks are within the same subject area. In the social sciences as well as the natural sciences, there was the view that information exchange and exchange of expertise was easier and more intense than through more informal mechanisms such as conferences. Overall, issues in terms of management and communication seem to be similar across social science and science areas.

Policy implications

On the basis of eight case studies, it is not possible to make firm policy recommendations. Nevertheless, these case studies do suggest various areas for future research and do raise a number of questions and suggestions relevant to policy. Key issues are:

- interaction
- scope economies
- information and knowledge-transfer, and
- managerial approach.

Benefits from networks are higher where there is high interaction and high scope economies while excessively tight management can be detrimental. Networks appear to facilitate greater knowledge transfer than more informal means of communication.

The case studies show that interaction and diversity are important elements in the existence and value-added of networks. The extent of interaction will depend on the nature of the project and the nature of the scientific area and will also depend on diversity among participants. This diversity does not need to be large; the case studies illustrate that substantial benefit can be gained from interaction and information exchange even where diversity arises from difference in individual experience expertise and approach rather than through interdisciplinarity. Interaction also depends on the structure and operation of a network and therefore it depends both on management and on participants.

One policy implication is that one way to try to assess ex ante the value and necessity of a network is to look at the likelihood of interaction and the basis and justification for this interaction. In addition, information and guidelines could usefully be provided for managers and for participants about ways of interacting and the potential benefits from interaction. On the basis of an ex ante evaluation, it may also be possible during the period of operation of a network to assess whether it is achieving the amount of interaction anticipated and the benefits anticipated from interaction. It is not possible to use crude indicators of interaction as evaluation measures as this may simply lead to a distortion in scientists' behaviour without actually affecting the research and collaborative network operation. Substantial information could, however, be obtained by evaluating the main aims of the network prior to its establishment. The network type can be identified as set out above and from that consideration given to whether the network is based largely on economies of scale, scope or externalities. Networks where little scope or interaction benefits are expected should be looked at especially carefully.

Overall this means that the benefits obtainable from diversity in networks need to be given more weight. Policy should focus more on economies of scope in networks rather than simply on economies of scale. As discussed above, when considering economies of scale, attention also needs to be given to the type of economies of scale – are they due to the common use of equipment, shared samples and data or to the overall size of the network and project. The implications of the two different types of scale economies are not the same.

As other studies have shown, the management and co-ordination role in networks is a central one. The case studies here suggest that it is important on the one hand that the management role is clear and agreed by all participants and on the other hand that management should be sufficiently flexible to allow participants sufficient autonomy. In particular, given the importance of interaction to most networks, interaction should not simply be between groups and the management, rather interaction needs to be between and across groups. This needs to some extent to be informal, it cannot all be managed tightly from the centre. Thus, while there is no one particular management style appropriate for networks, most networks will need to give some consideration to how interaction is to be facilitated and what the implications of this are to managerial style. The importance of the management and co-ordination roles suggests that guidelines, information and advice could usefully be given at the start of the network. On the one hand, clarity of structure and aims of the network is important both for co-ordinators and participants; on the other hand, networks need to have sufficient flexibility to be able to change and adapt their structures, and to develop their aims, over time.

The two main alternatives to networks are to concentrate researchers and resources in a single facility or to put the emphasis on more informal means of interaction and communication. The case studies suggest that networks offer substantially more flexibility than concentration in a single facility and benefits from diversity are less likely to be undermined in a network since participants remained in their different environments. On the other hand, networks appear to have a rather strong advantage over more informal methods of communication, particular that, due to learning, trust and common purpose, participants in networks experience more intense information and knowledge transfer than they would through more informal methods of interaction and find networks more important as a way of keeping up to date in a specialised area of a subject. These results appear to be similar to the observation and analysis of informal know-how trading among firms (von Hippel 1987).

National and EU science policy

There are a series of issues raised by this study for the relationship between national and EU science policy. Although the study found no major differences between national and European networks, this does not imply that there are or should be no major differences between UK and European science policy. There is no justification for the EU funding individual projects in individual institutes unless scale economies are so large that the project involves focusing researchers from across countries into one institute. Otherwise, funding of single institutes and single projects remains the responsibility of national governments. Scientists tend to participate in a number of networks and in order to contribute to these networks they need to have a strong base in their own area and in terms of their own knowledge and expertise. In most cases, this means they all need to be carrying out some part of their work on single projects within their own institute. In subjects that are increasingly interdisciplinary, this may be less so. The policy question for national governments is the relative emphasis that should be placed on single institute work, national collaboration and international collaboration. The policy question for both national governments and the European Commission is their relative roles with respect to international science and international collaboration.

The question of subsidiarity in science policy is a complex one for two main reasons: the international nature of science, and the importance of scope economies in underpinning benefits from collaboration (see chapter one for a fuller discussion of these issues). Where scope economies are important, there is no simple distinction between those cases appropriate for national policy and those appropriate for EU policy, as there is for scale economies. Where science is international, then there is a role for national and not only EU policy in promoting this. For example, where international collaboration between a small number of people or institutes is appropriate, or where collaboration outside the EU is relevant, then national science policies may be the appropriate institutional structure. Where both national and EU science policies are concerned with international science, then the nature of subsidiarity is complex and a variety of institutional structures may be needed.

There are a number of further issues about the inter-relationships between UK and EU science policy. There are unintended as well as intended interactions between the two, as discussed in chapter one. EU science policy may change the distribution of funds for science within the UK with knock-on implications for the behaviour of the UK funding bodies. It may act to change the research areas and focus of researchers in the UK and it will affect their international links. As industry receives substantial amounts of EU funding, EU policy also affects relationships between UK industry and UK universities. It may be that in some cases where industry would previously have funded university research, it now suggests that industry and universities jointly go to the EU for funding. This will potentially change the nature of the research, including encouraging international collaboration, and it will change the distribution of funds not only across areas but also between firms and universities.

Conclusion

Co-operative research networks are increasingly common forms of organisation of scientific activity. Network structures will affect research both within the networks and the overall developments of scientific fields. It is, therefore, important to increase our understanding of the operation of these networks. This study aims to contribute to this understanding. Key issues in the analysis of networks arising from the current study are: interaction and diversity, scope economies, information and knowledge transfer, and managerial approach. The development of international networks and international interaction also raises major questions about appropriate policy structures. In particular, many issues concerning the appropriate roles of national and EU science policies remain to be resolved.

References

Baumol, W. (1992) 'Horizontal Collusion and Innovation', *The Economic Journal,* Vol.102, pp.129-137

CEC (1992) 'Evaluation of the Second Framework Programme for Research and Technological Development' Commission of the European Communities, SEC (92) 675 Final, Brussels

Chesnais, F. (1988) 'Technical Co-operation Agreements Between Firms' *STI Review,* No.4, OECD, Paris

Cohen, L.D. (1994) 'When can Government Subsidize Research Joint Ventures? Politics, Economics and Limits to Technology Policy', *American Economic Review,* Papers and Proceedings, May, pp.159-163

Crane, D. (1972) *Invisible Colleges,* University of Chicago Press, Chicago

Dasgupta, P. and E. Maskin (1987) 'The Simple Economics of Research Portfolios', *Economic Journal,* Vol.97, pp.581-95

de Meyer, A. (1992) 'Management of International R&D Operations' in Granstrand, L; Håkanson, L., Sj Ölander, S. (eds) *Management and International Business: internationalisation of R&D technology,* John Wiley, Chichester

Dickson, K. Lawton Smith, H. and Lloyd Smith, S. (1991) 'Bridge over Troubled Waters? Problems and Opportunities in Interfirm Research Collaboration', *Technology Analysis and Strategic Management,* Vol.3, No.2, pp.143-156

Edgerton, D. and Hughes, K. (1993) 'British Science Policy in the 1990s: Technology and the Market', mimeo

Franklin, M.N. (1988) *The Community of Science in Europe,* Gower, Aldershot

Freeman, C. (1987) *Technology Policy and Economic Performance,* Pinter, London

Fusfeld, H. and C. Haklisch (1985) 'Co-operative R&D for Competitors', *Harvard Business Review,* November-December, pp.60-76

Geroski, P. (1992) 'Antitrust Policy Towards Co-operative R&D Ventures', *Oxford Review of Economic Policy,* Vol.9., No.2, pp.58-71

Geroski, P. and A. Jacquemin (1985) 'Industrial Change, Barriers to Mobility and European Industrial Policy', *Economic Policy,* Vol.1, pp.170-218

Grossman, G. and C. Shapiro (1987) 'Dynamic R&D Competition', *Economic Journal,* June, pp.372-86

Hagedoorn, J, and Schakenraad, J. (1993) 'Strategic Technology Partnering and International Corporate Strategies' in Hughes, K. (ed) *European Competitiveness,* Cambridge University Press, Cambridge

Hingel, A. J. (1993) 'Note on "A New Model of European Development", Innovation, Technological Development and Network-led Integration' Commission of the European Communities, FAST

House of Commons (1993) 'The Routes Through Which the Science Base is Translated into Innovative and Competitive Technology', Minutes of Evidence, 8 December, Science and Technology Committee, London HMSO

House of Commons (1994) *The Routes Through Which the Science Base is Translated into Innovative and Competitive Technology,* First Report, Science and Technology Committee, London HMSO

House of Lords (1990) *A Community Framework for R&D,* Select Committee on the European Communities, HL Paper 66, London, HMSO

House of Lords (1993a) *European Community Fourth Framework Programme for R&D,* Select Committee on Science and Technology, HL Paper 5, London, HMSO

House of Lords (1993b) *Priorities for the Science Base,* Report, Select Committee on Science and Technology, HL paper 12-I, London, HMSO

Hughes, K. (1989) 'The Changing Dynamics of International Technological Competition' in Audretsch, D., Sleuwaegen, L., and Yamawaki, H. (eds) *The Convergence of International and Domestic Markets,* North Holland, Amsterdam

Hughes, K. (1992) 'Technology and International Competitiveness', *International Review of Applied Economics,* Vol.6, No.21 pp.166-183

Imai, K. I. and H. Itami (1984) 'Interpenetration of Organisation and Market', *International Journal of Industrial Organisation,* Vol.2, pp.285-310

IMPACT-UK (1993) 'The Impact of European Community Policies for Research and Technological Development upon Science and Technology in the United Kingdom' by Georghiou, L., Stein, J. A., Jones, M., Senker, J., Pifer, M., Cameron, H., Nedeva, M., Yates, J., Boden, M. Report for DGXII of the

Commission of the European Communities and the UK Office of Science and Technology

Jacquemin, A. (1988) 'Co-operative Agreements in R&D and European Antitrust Policy', *European Economic Review,* 32, pp.551-560

Katz, M. (1986) 'An Analysis of Co-operative Research and Development', *Rand Journal of Economics,* Vol.17, No.4, pp.527-43

Katz, M. and Ordover, J. (1990) 'R&D Co-operation and Competition', in Baily, N. and Winston, C. (eds) *Microeconomics,* Brookings Papers on Economic Activity, Brookings Institution, Washington, D.C.

Leclerc, M. Okubo, Y., Frigoletto, L. and Miquel, J-F, (1992) 'Scientific Co-operation between Canada and the European Community', *Science and Public Policy,* Vol.19, No.1, pp.15-24

Lei, D., and Slocum Jr., J. (1992) 'Global Strategy, Competence-Building and Strategic Alliances', *California Management Review,* Fall

Link, A. and Tassey, G. (eds) (1989) *Co-operative Research and Development: the Industry-University-Government Relationship,* Kluwer Academic Publishers, Boston

Mowery, D. (1987) *Alliance Politics and Economics,* AEI/Ballinger

Mytelka, L. K. (1990) 'New modes of international competition: the case of strategic partnering in R&D', *Science and Public Policy,* Vol.17, No.5, pp.296-302

Nelkin, D. and R. Nelson (1986) 'commentary' in *New Alliances and Partnership in American Science and Engineering,* National Academy Press, Washington D.C

Nelson, R. (1984) *High Technology Policies – A Five Nation Comparison,* AEI, Washington D.C

OECD (1992) *Technology and the Economy – the key relationships,* OECD, Paris

Ohmae, K. (1985) *Triad Power,* The Free Press, New York

Ordover, J. and R. Willig (1985) 'Anti-Trust for High Technology Industries', *Journal of Law and Economics,* Vol.28, pp.311-33

Patel, P. and Pavitt, K. (1987) 'Is Western Europe Losing the Technological Race?', *Research Policy,* Vol.16, Nos 2-4, pp.59-86

Patel, P. and Pavitt, K. (1991) 'Europe's Technological Performance' in Freeman, C., Sharp, M., and Walker, W. (eds) *Technology and the Future of Europe,* London, Pinter

Reich, R. and E. Mankin (1986) 'Joint Ventures with Japan Give Away Our Future', *Harvard Business Review,* March-April, pp.78-86

Rosenberg, N. (1982) *Inside the Black Box: Technology and Economics,* Ch.12, Cambridge University Press, Cambridge

Rosenberg, N. and C. Frischtak (eds) (1985) *International Technology Transfer,* Praeger, New York

Science Policy White paper (1993) *Realising our Potential: a strategy for science, engineering and technology,* Cm 2250, London, HMSO

Sharp, M. (1991) 'The Single Market and European Technology Policies', in Freeman, C., Sharp, M., and Walker, W., *Technology and the Future of Europe,* Pinter Publishers, London

Sharp, M. and Shearman, C. (1987) *European Technological Collaboration,* Chatham House Papers 36, The Royal Institute of International Affairs, Routledge and Kegan Paul, London

Shepherd, G., Duchene, F. and C. Saunders (eds) (1983) *Europe's Industries,* Frances Pinter, London

Teece, D. (1977) 'Technology Transfer by Multinational Firms: The Resource Cost of Transferring Technological Know-How', *Economic Journal,* Vol.87, pp.242-61

Teece, D. (1981) 'The Market For Know-How and the Efficient International Transfer of Technology', *Annals,* No.458, pp.81-96

Vickers, J. (1985) 'Pre-emptive Patenting, Joint Ventures and the Persistence of Oligopoly', *International Journal of Industrial Organisation,* Vol.3, No.3, pp.261-73

Von Hippel, E. (1987) 'Co-operation between rivals: Informal know-how trading', *Research Policy,* Vol.16, pp.291-302